To Robin —

R E Craft

Lisa Honebrink

By the turn of the 21st century, Phoenix was served by a multitude of broadcast television signals, not to mention all of those that could be received through cable or direct broadcast satellite. The many local stations placed their transmitter buildings and towers on top of South Mountain. The antenna farm now located there presents an impressive sight of a forest of towers. (Photograph by John E. Craft.)

ON THE COVER: Celebrating its Western roots, Phoenix was noted for hosting rodeo parades in the 1950s. KPHO-TV Channel 5 covered this parade on both film and television. The film cameras on the Ford station wagon provided a permanent record of the event, while the television camera and microwave dish on top of the remote van enabled broadcast of the spectacle live to the Phoenix television audiences. Ralph Painter is operating the large 16mm-film camera at center. (Courtesy of KPHO-TV.)

Phoenix Television
Images of America

Dr. John E. Craft and Lisa Honebrink

Copyright © 2019 by Dr. John E. Craft and Lisa Honebrink
ISBN 978-1-4671-0279-7

Published by Arcadia Publishing
Charleston, South Carolina

Printed in the United States of America

Library of Congress Control Number: 2018955593

For all general information, please contact Arcadia Publishing:
Telephone 843-853-2070
Fax 843-853-0044
E-mail sales@arcadiapublishing.com
For customer service and orders:
Toll-Free 1-888-313-2665

Visit us on the Internet at www.arcadiapublishing.com

Contents

Acknowledgments		6
Introduction		7
1.	KPHO-TV Channel 5	11
2.	Wallace and Ladmo	33
3.	KTYL-TV, KVAR-TV, KTAR-TV, and KPNX-TV Channel 12	45
4.	KOOL-TV, KTSP-TV, and KSAZ-TV Channel 10	57
5.	KTVK-TV Channel 3	79
6.	KAET-TV Channel 8	97
7.	KNXV-TV Channel 15	109
8.	Other Phoenix Television Stations	117
The House of Broadcasting Inc.		127

ACKNOWLEDGMENTS

The authors would like to thank all of those who have assisted in making this work possible. Certainly, the stations and individuals who provided photographs of past television stations, on-air personalities, behind-the-scenes artists and technicians, and historical programs deserve our thanks and are duly acknowledged following the caption describing each photograph, but others also have contributed to this work. For example, to conduct the research necessary to complete this volume, the authors talked extensively with people who devoted their careers to bringing television and its entertainment and informational programming to the residents of Phoenix.

The authors also would like to thank the members of the board of the House of Broadcasting Inc. This 501c3 organization is dedicated to preserving the history of broadcasting in the Phoenix market through the display of equipment and programming artifacts from local radio and television stations. Special thanks go to the founder of the House of Broadcasting, Mary Morrison, and one of its greatest benefactors, Jack Clifford.

Most importantly, the authors would like to thank the pioneers of Phoenix broadcasting, as well as all those professionals who have been working diligently over the years to bring information and entertainment to the Phoenix market through what has become the most pervasive mass communication tool yet invented. According to Noll & Associates, *Media Executive Summary* Volume 24, No. 28, more than 96 percent of all the homes in the United States have at least one television set, and Phoenix is no exception.

Over the past 70 years, the dedicated broadcasters of Phoenix have brought a vast array of quality entertainment and needed information to us, the citizens of this region. Television, and its content, have done much to shape our society and we thank our Phoenix broadcasters past and present.

We also thank editors Angel Hisnanick and Lindsey Givens of Arcadia Publishing for their assistance in the completion of this book.

Introduction

Before television came to town, Phoenix of the early 1940s was not a large city. Located in the heart of the fertile lands of the Salt River Valley, it was blessed with abundant sunlight and water flowing through the canals of the Salt River Project. These were the ingredients that allowed the desert to bloom with fields of cotton and alfalfa as well as endless groves of orange trees. However, before World War II, living in the Valley was for the hardy only—especially when the summer temperatures might top out at 115 degrees. At the same time, the high-pressure weather patterns that allowed those sunny days to prevail created the perfect environment for flying. Subsequently, the Army created aviation training bases such as Luke Air Force Base, Williams Field, and Thunderbird. Those bases brought many people to the Southwest and introduced them to the delights of the dry desert air and the endless sunlight.

By the late 1940s, the economy was good; men and women had returned from service in World War II, wartime rationing was over, and the demand for new homes, new cars, and new appliances was rampant. When the hostilities ended, returning soldiers, retired business folk, and electronic manufacturing giants all wanted to live in the land of the perpetual sun, thus exploding the population into what is now the country's fifth-largest city.

Developers such as John F. Long and Del Webb forged their contributions to the growth of the Phoenix area. Long provided inexpensive housing that the returning service personnel needed for their families, building Maryvale in Phoenix as one of the nation's first master-planned and affordable communities. Later, Del Webb built Sun City, a planned community complete with resort amenities, to attract the recently retired to Phoenix.

Early electronic manufacturers, such as Motorola, provided the needed employment for the city to grow. Companies such as Goettl and Shasta provided the air conditioning and swimming pools that made living in the desert seem like one long summer-resort vacation, while Basha's provided the groceries for the backyard barbecues. That said, some new residents from the more populated East and Midwest also felt the isolation of living in the relatively new and sparsely settled state, cut off from the large population centers of New York City, Chicago, and Los Angeles by the formidable Sonoran Desert. Welcome as a cold drink on a Phoenix summer day, the technological wonders of network radio, Hollywood motion pictures, and then, early broadcast television, allowed the new desert immigrants to feel a cultural kinship with their fellow Americans.

Due to technological development (primarily with radar) during the war years, a new home entertainment and information medium—television—was poised to revolutionize the way Americans received their news, reacted to dramas, laughed at comedies, and cried during the soap operas. Television incorporated the popularity of radio and added pictures. Television was to change the way we interacted as a family, related to our communities, and even how we ate our dinner.

During the World War II years, motion pictures provided escapist entertainment to the public, and radio brought eagerly sought-after news directly into the home. Edward R. Murrow's voice in your living room, describing the bombs dropping on London, half a world away, even while it was happening, seemed like a miracle; yet even while sending a voice instantly through thousands of miles of air, other inventors were dreaming of broadcasting live pictures that moved along with the sound.

Many experimented with what was to become television. Initially, a mechanical camera system was envisioned to break the elements of the picture into small bits of varying intensity of light that could be turned into a signal in the electromagnetic spectrum. The signal was sent through the airwaves to a receiver at a remote location. The receiver, using a similar scanning disk, reassembled the picture to give a tiny low-resolution image that vaguely represented what was originally transmitted. By the late 1920s, hobbyists experimenting with the emerging field of electricity could order many of the parts needed to build their own television receivers. Of course, there was nothing to watch and the system produced only the crudest of outlines of the images. Probably the biggest disadvantage of the mechanical system was the large disk rapidly spinning like a circular saw in the living room. The mechanical system was doomed to fail.

Luckily, others were working on television systems that scanned the image electronically. A farm boy from Idaho, Philo T. Farnsworth, moved to San Francisco, established a small workshop, and began to work out his vision of a method of scanning and dissecting a picture for transmission without using any moving parts. His system, once patented, allowed Farnsworth to become known as the "father of television."

The massive electronic manufacturing company Radio Corporation of America (RCA) was also very interested in developing an electronic television system. RCA president David Sarnoff hired the Russian-born scientist Vladimir Zworykin to develop a workable system using many of Farnsworth's discoveries. By 1939, RCA had developed an electronic television system that was refined enough to be presented to the public at the "World of Tomorrow"–themed World's Fair in New York City. The demonstration was a tremendous success and even included the fair's opening address by Pres. Franklin Roosevelt. Roosevelt was seen live on the 200 television receivers located throughout the fair's grounds. The public was given a glimmer of what they might "see" in the future.

Then World War II intervened, and the few experimental television stations in existence at that time went off the air for the duration. However, by 1944, several experimental stations, primarily in New York City, began transmitting again on a very limited basis. During the next year, the Federal Communications Commission (FCC) allocated spectrum space for new television channels and by January 1, 1946, *Broadcasting Yearbook* listed nine television stations either on the air or with permits to construct a station. As with radio, the owners of these stations realized the economics of sharing programs and began to interconnect the television stations with co-axial cable links. Thus, the television networks were born.

RCA owned National Broadcasting Company (NBC), William S. Paley owned Columbia Broadcasting System (CBS), and Allen B. DuMont, an early manufacturer of television receivers, owned the DuMont television network. Although NBC and CBS had owned and operated very profitable radio networks for years, DuMont is often credited with the first interconnection of television stations to form a network, when it connected the stations that it owned in Washington, DC, and New York City. American Broadcasting Company (ABC) was formed after RCA was required to divest one of its two radio networks in 1943, and was a slow entrant into television. The Mutual Broadcasting System, a popular radio network, never entered the television business.

The early television stations and networks were nothing without the television receiving set. Early inventors such as Charles F. Jenkins, Philo Farnsworth, Allen B. DuMont, and Earl "Madman" Muntz all established manufacturing firms to market early television sets, as did large electronics companies such as Philco, Zenith, General Electric, and RCA. Early sets were available as far back as the 1920s. Some used mechanical systems for viewing, but others advanced to cathode ray tubes for direct electronic viewing. Viewing surfaces as large as 1.5 inches were enhanced with a magnifying glass placed in front of the screen. While the electronics of the sets were basic at best, the large stylish cabinets were of the finest furniture, designed to blend into the poshest of upscale living rooms. The furniture quality had to be the best, because only the very wealthy could afford a television set, which might sell for more than half the price of a new automobile.

As the technology advanced during the late 1940s and early 1950s, the size of the viewing surface increased and a reflecting mirror and magnifying glass were no longer needed. More importantly, the cost of the receiver began to decrease, thus allowing more and more families to own the latest home entertainment and information device. By 1948, the average cost of a set was about $400, which was expensive by the standards of the time but within reason for many Americans. When network programs were broadcast on a more regular schedule in 1948, only two percent of American homes had a receiver, and many saw television in bars, department stores, and appliance and radio shops. Interest in television became so great that in only eight years, the percentage of homes with television sets had reached 70 percent. Never before had a technology been adopted so quickly. The key to the fantastic growth in television set ownership was the programming that each of the networks provided to its local affiliated stations.

The public's desire for television was such that many owners of radio stations, newspapers, and hardware stores—or just citizens interested in the new technology—applied to the FCC for a license to operate a television station, hoping to cash in on what some thought might become a new gold mine. So many applied for licenses that the FCC quickly ran out of frequencies, or channels, on which the stations could operate. In 1948, the FCC put a freeze on issuing licenses until a new chart of frequencies could be established. In 1952, the FCC once again offered licenses for television stations, but many were on the newly formed UHF band of channels 14 through 83. Channel 1 disappeared at that time. Between 1948 and 1952, many cities had only one television station, and that station could pick and choose from one of the four television networks to fill its programming schedule.

That program schedule was very different from the fare offered on the numerous broadcast, cable, and streaming networks available today. Early programming on the four commercial networks was live for the stations interconnected by AT&T long lines and on grainy black and white 16mm film (called kinescopes) on stations located in the smaller remote cities. Often, the kinescope copies were sent to the local stations many days or weeks after the original broadcast and were delivered by Greyhound Bus Lines.

Before 1953, when ABC merged with Paramount theaters, most network programming broadcast only a few hours each evening, originated from New York City, and was influenced by Broadway stage production techniques. It was only later that many of the network shows moved to the West Coast and copied the motion picture production techniques developed there.

Television quickly became the medium of choice for the American public to get their entertainment, news, and even commercials, because it came directly into the living room with moving pictures, dialogue, and music. It provided comedy, drama, action adventure, soap operas, sports, movies, and news. It took the viewer to many exotic places in the world—while they could sprawl in air-conditioned comfort in a recliner and enjoy an icy Coca-Cola. The picture on the tiny screen in the mahogany box in the corner of the family room may have been fuzzy and only in black and white, but it was in your home, and you could brag to your friends that you had one of the newest entertainment devices that money could buy. The best part of television was that it was free—that is, if you could afford the TV set. A nice television receiver packaged with an AM/FM radio and a phonograph, housed in a handsome piece of furniture, might cost up to 10 percent of the price of a brand new, three-bedroom ranch house in Maryvale. And yet the demand for television was overwhelming. According to the *World Book Encyclopedia*, there were fewer than 10,000 television receivers in the United States in 1945, but that number had bloomed to over six million just five years later.

Which of us, of a certain age, can forget *I Love Lucy*, *Death Valley Days*, *Toast of the Town*, and *To Tell the Truth*? But early on, before the networks had a full schedule of soap operas, game shows, movies, and variety shows, the local stations had to fill their broadcast time with locally produced programs or rented movies and cartoons. The alternative was to simply broadcast a test pattern and tone for hours during the daytime. Early on, many stations did this—and many viewers watched.

Television stations in Phoenix and across the country were pioneering new forms of information, entertainment, and commercials as they filled the broadcast day. Many local programs mimicked what was seen on the networks. News programs often were simply a newsreader sitting at a desk in front of a gray drape and reading from a script torn from a newswire machine. News from the local station looked the same as that coming from the network. While many original programs were produced for the local television audience, similar programs were found in the schedule of almost every local station in the country.

Almost every station had a kid's program. Typically, that consisted of a station employee dressing as a clown and introducing cartoons. Chicago had *Bozo the Clown* and Phoenix had *Wallace and Ladmo*. The latter, broadcast on KPHO-TV, became one of—if not the—longest-running children's show in the country with the same cast, celebrating its 35th anniversary in 1989. *The Wallace and Ladmo Show* became the best-known television program in Phoenix and entertained generations of adults as well as children.

Most stations also hosted a women's program. This filled time in the mid-morning hours with tips for homemakers and hours of cooking demonstrations. Many of these shows also provided an interview segment in which local newsmakers could make a pitch to their local public. Channel 12 offered a women's/news program titled *Today in Arizona* hosted by Diane Kalas. Later, Channel 5 featured Rita Davenport in *Open House*.

Local stations found that they could rent "B" grade movies of the 1930s and 1940s for very little money. These became the programs that often filled the late afternoon hours and definitely the late-night time period. Sometimes, a local host described the upcoming scenes of the film, and always, the movie was interrupted every few minutes with a series of commercials.

Another common program was the after-school dance party. Local teens were invited into the studio to dance to the tunes of local bands. The local dance programs followed the format developed by a Philadelphia station but usually had a somewhat inferior local band and no Dick Clark. In Phoenix, however, KPHO-TV offered *Teen Beat*, hosted by KRIZ radio deejay and program director of the year Pat McMahon. In addition to the local Salt River Navy Band with Mike Condello, the show was able to book national artists such as Bobby Vinton, Tuesday Weld, Johnny Cash, Frankie Avalon, and the Everly Brothers. Phoenix TV was on the map, although many of these artists were in their salad days.

In the very early days of television, local commercials were performed by the program host in the studio live in front of a camera. Generally, the sets were very simple, the graphics and music were almost nonexistent, and the mistakes were very common and often hilarious. Remember Acquanetta pushing her husband's Mercury cars or an early Tex Ernhardt and his "No Bull" commercials? Sales pitches were simpler but accomplished the goals of the advertiser and eventually brought in large profits for the stations.

Today, with the domination of three major networks long gone and many viewers live streaming shows on-demand on hand-held devices, it is still hard to imagine life without television.

This book will describe how family-owned television stations were established in the Phoenix market—allowing viewers to mitigate their isolation and feel a cultural kinship with their fellow Americans—and will allow the reader to bask in reminiscence of bygone favorite local programs.

One
KPHO-TV Channel 5

Phoenix got its first view of television when the new communication medium was still very young. In 1949, John C. Mullins and partners, owners of KPHO radio, put Arizona's first television station on the air. KPHO-TV broadcast on channel 5 from a tall steel tower attached to the Westward Ho Hotel on Central Avenue. As the first station in Phoenix, it was able to choose programs from any of the four networks then in existence: NBC, CBS, ABC, and DuMont. Even with four networks, programming choices in those days were limited, and KPHO-TV, like most network affiliates, produced many local programs as well as live local commercials. Early KPHO-produced television programs included *The Lew King Ranger Show*, *The Goldust Charlie Show*, and Arizona's first TV newscast with Jack Murphy.

Perhaps the best known local program on KPHO-TV was *The Wallace and Ladmo Show*. Airing five days a week from 1954 to 1989, it made history as one of the longest running locally produced children's programs in the country.

KPHO-TV's studio was located on First Avenue just west of the Westward Ho. Now aptly named First Studio, it is still used as a TV production facility today.

In 1952, the Meredith Corporation, owner of *Good Housekeeping* magazine, purchased KPHO-TV. Within a year, KPHO-TV Channel 5 began to lose network programming choices as stations went on the air on channel 12, channel 10, and channel 3 and entered into affiliation contracts with the major networks. When the DuMont network went out of business in 1956, KPHO-TV lost its major network programming source. For the next 38 years, Channel 5 operated as a successful independent station by broadcasting off-network reruns, old movies, and locally produced programs.

In 1994, in a wide-ranging affiliation swap, KPHO-TV regained the CBS network contract. In 2014, the Meredith Corporation acquired independent KTVK-TV Channel 3 and moved KPHO-TV from its location of more than 40 years (at Indian School and Black Canyon Highway) to Channel 3's larger building on North Seventh Avenue. Meredith operates both channels 3 and 5, using many of the same employees to run both stations.

Channel 5's original building was located on First Avenue next to the Westward Ho Hotel. The tower on top of the hotel was the original home of the broadcasting antenna for the station, but by the late 1960s, when this picture was taken, all Phoenix transmission towers were located on South Mountain. (Courtesy of Sharon L. Kelley.)

When located near the Westward Ho Hotel in downtown Phoenix, Channel 5 promoted its upcoming attractions on a large movie theater–style marquee, which often included the advertiser's name to provide extra value to those buying TV commercials. (Courtesy of KPHO-TV.)

The Goldust Charlie Show was one of the first kids' shows on pioneering Phoenix station KPHO-TV Channel 5. Ken Kennedy played the role of Goldust Charlie, who was the proprietor of a general store in the Old West. Bill Thompson, later of *Wallace and Ladmo* fame, got his start on the floor crew of this program. (Courtesy of Sharon L. Kelley.)

KPHO bought advertising in the November-December 1954 *TV Views* guide, to support the publication as well as to remind its audience of the times that the popular shows were scheduled. *The Goldust Charlie Show* was one of several local children's programs of the early 1950s. (Courtesy of Robert T. Martin.)

Although this promotional photograph might suggest an action-filled Western, the *Lew King Rangers Show* (hosted by cowboy radio star Lew King) was a weekly kid's talent show, which moved to KPHO-TV in 1950. Weekly auditions decided which child would get to perform. Wayne Newton debuted on the show, along with other future stars such as Tanya Tucker, Marty Robbins, and Lynda Carter. (Courtesy of House of Broadcasting Inc.)

KPHO-TV, Arizona's first television station, was launched on December 4, 1949. Jack Murphy, Arizona's first television anchorman, is shown here with his floor crew. He continued as Channel 5's first news anchor, news director, and executive producer. (Courtesy of KPHO-TV.)

Pictured is actor George Cleveland (left) with Jack Murphy, KPHO-TV's news anchorman. Cleveland is best remembered as Gramps on the original 1954 *Lassie* TV series. (Courtesy of House of Broadcasting Inc.)

KPHO-TV Channel 5 broadcast many live telethons throughout its history. Here, a broadcast before a live audience helps to raise money for the March of Dimes. This production took place at Channel 5's first studio at 631 North First Avenue. The original DuMont cameras have "Inky Dink" lights mounted on the front to add sparkle to the eyes of talent on close-ups. Cameras of the time required very bright light to produce viewable pictures. (Courtesy of KPHO-TV.)

In the early 1960s, Marge Condon hosted *Open House* on KPHO-TV. The station, like many others, constructed a full kitchen in the studio so it could broadcast culinary demonstrations. From the earliest days, the station broadcast cooking programs. A popular show airing five days a week in the early 1950s was called *Cook's Corner*. (Courtesy of KPHO-TV.)

In the early days, in addition to carrying programs from all four networks, KPHO-TV provided remote coverage from locations outside the studio. Here, the station originates coverage from a Del Webb construction site. The camera operator on top of the trailer transmits the signal to the microwave dish affixed to the scaffolding on the truck. That signal is then sent to the station's tower on the Westward Ho Hotel. (Courtesy of KPHO-TV.)

For television, the 1950s and 1960s were years of invention. Stations were game to try almost anything to get a new, unusual shot. Pictured is KPHO-TV's homemade underwater camera box, allowing Channel 5 to air Arizona's first underwater television shots. Given the high voltage required to operate the cameras of that time, the crew is showing great faith in the water-tight construction of their creation. (Courtesy of KPHO-TV.)

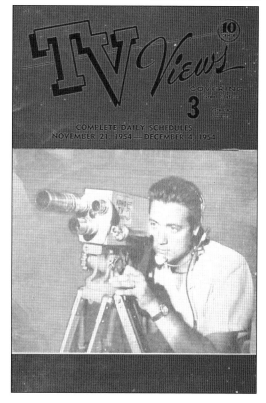

The cover photograph of this 1954 *TV Views* features Channel 5's foray into the latest technology. The mini video camera was designed to allow video to be shot at remote locations with a minimum of equipment and personnel. The camera was to compete directly with the 16mm-film camera. However, it would be another 20 years before a practical minicam was widely adopted by television news departments. (Courtesy of Robert T. Martin.)

Channel 5's Bob Martin (right) selects the correct lens on the DuMont camera to frame a close-up of the unidentified talent. Martin was vice president of programming at KPHO from 1952 to 1962, when he moved to KOOL-TV, heading that station's programming, promotion, and operations until 1982. Martin started his career at the CBS Television Network in 1946 at the tender age of 16. (Courtesy of Robert T. Martin.)

Long before computers were available to local television stations, all artwork that appeared in programs and commercials had to be created by hand. Most stations employed graphic artists to create the visual images, which were then turned into 35mm slides for broadcast. Pictured is staff artist Emmett Lancaster working out of the original KPHO offices in downtown Phoenix. (Courtesy of Sharon L. Kelley.)

Programs recorded on 16mm film were stored on a metal reel. One hundred feet of film was needed for about four minutes of content with the projector running at 24 frames per second. Film not properly rewound on a reel could be a problem. (Courtesy of Randy Murray.)

Local television news sets in the early days were not sophisticated, and the weather maps definitely were not high tech. The jumble of papers and scripts on the table indicates that KPHO-TV had not installed teleprompters on their cameras yet. (Courtesy of KPHO-TV.)

Dialing for Dollars was a popular local segment on KPHO-TV in the 1950s. Network game shows that gave away money and prizes included *Strike it Rich*, *Beat the Clock*, and *Pick the Winner*. (Courtesy of Randy Murray.)

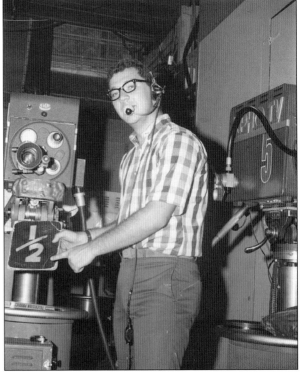

A floor manager stands next to the camera and gives cues to the on-air talent as to which camera to look at and the amount of time left in a segment. The floor manager communicates with the director through a headset and informs the talent that he/she has 30 seconds left to the end of the segment. (Courtesy of Randy Murray.)

Several engineers were required in the control room during the broadcast. In this photograph, an audio engineer selects the correct microphone and controls the volume by turning the knob on the audio board. The VU (volume unit) meter in the center of the board allows him to see the output of the console. (Courtesy of Randy Murray.)

This engineer is monitoring the quality (brightness and contrast) of the signal that is coming from each studio camera. As the light levels changed with each shot, the engineer would compensate by adjusting the output of the camera. He checked the signal output of the camera by viewing the pattern on the oscilloscope at far right. (Courtesy of Randy Murray.)

Engineer John Foley prepares to start a film projector at the command of the director, who is communicating through the headsets to all the technical and studio crew. The three-ring binder in front of Foley contains the script for the program. (Courtesy of Sharon L. Kelley.)

In 1961, the Phoenix chapter of the National Academy of Television Arts and Sciences awarded its first Governor's Award to KPHO-TV. The Phoenix chapter, now the Rocky Mountain Southwest chapter, was just the third chapter in the nation to be organized after the established chapters in New York and Los Angeles. (Courtesy of KPHO-TV.)

A practical color TV system was introduced by RCA in 1953, and by the mid-1960s, most network primetime programs were broadcast in color. Shortly after that time, many local television stations began to broadcast local news in color. KPHO-TV's first color studio camera, shown here, was an RCA model TK-41. This camera freed viewers from the limitations of "only" black-and-white programs. (Courtesy of Randy Murray.)

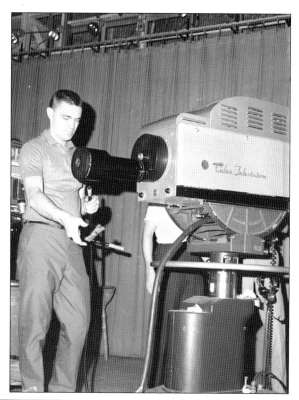

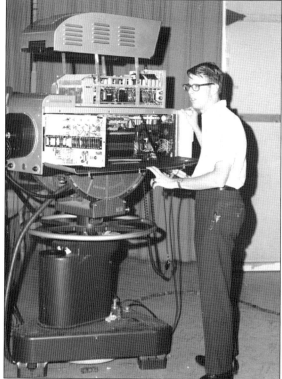

Television equipment of the 1950s and 1960s was very temperamental and required daily maintenance. Engineers were constantly testing, repairing, and adjusting the various electronic components that were required to create and transmit the television picture. (Courtesy of Randy Murray.)

The first locally produced color program in Phoenix was *The Wallace and Ladmo Show*. KPHO-TV beat KOOL-TV in broadcasting a local program in color by just two hours; on the same day, KOOL-TV debuted color on its evening news. This RCA TK-41 camera is focused on Ladmo and Mike Condello. (Courtesy of KPHO-TV.)

KPHO-TV's Kathy Nolan is pointing out the highs and lows in this c. 1960s photograph, which represents the "weather girl" craze of that era. Women were accepted as weathercasters as long as the focus was kept on clothing, hairstyle, and looks. "Weather girl" is a term that speaks volumes about the differences in status between women and their male counterparts, who were called weather men. (Courtesy of House of Broadcasting Inc.)

In a previous era, the male news anchor would also give the weather forecast. Here, news director and anchor Jack Murphy demonstrates his weathercasting skills to guests in the KPHO-TV studio. (Courtesy of KPHO-TV.)

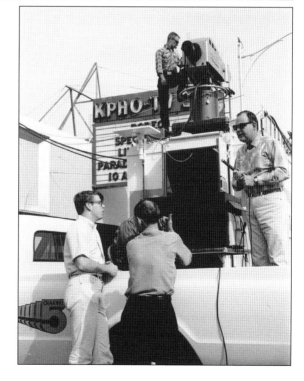

In this photograph, KPHO-TV engineering staff is preparing to cover a remote event outside of their studio. They have installed their new RCA TK-41 color camera on top of the remote van—no small task given the cumbersome weight of the camera and pedestal. (Courtesy of KPHO-TV.)

Channel 5 is preparing to televise a rodeo parade in downtown Phoenix in the 1960s. The very heavy color camera has been hoisted atop the remote truck in order to get a clear shot of the parade. The camera operator appears to be dressed appropriately. The announcer will sit at the card table in the back of the pickup truck. (Courtesy of Randy Murray.)

In 1959, KPHO-TV received its first videotape recorder. This first-generation RCA TRT-1 machine used two-inch magnetic tape and had six equipment racks to hold all the vacuum tubes, capacitors, and resistors required to make it work. The advantage of videotape was that the television images could be recorded and played back immediately. (Courtesy of Sharon L. Kelley.)

Engineer Steve Ulery adjusts the audio levels coming from audio sources such as film projectors, video recorders, microphones, or audio cartridge machines during a local production. The meters at the top of the audio board give the operator a visual indication of the strength of the audio signal. (Courtesy of Sharon L. Kelley.)

KPHO-TV has just received a new model two-inch videotape recorder. The large reel of two-inch-wide magnetic tape could hold enough information to save and play back a one-hour television program. The factory representative is explaining its features to Channel 5 engineers. (Courtesy of KPHO-TV.)

KPHO-TV's production control room in the 1980s included separate video monitors for each camera, videotape recorder, film chain, or character generator. The large monitors near the center provided preview and program output from the video switcher. The three larger monitors near the top allowed the director to preview special effects, such as graphics inserted into a picture. (Courtesy of KPHO-TV.)

In 1976, Sharon Kelley started her career in television by working on the floor crew at KPHO-TV. She went on to become the director of *The Wallace and Ladmo Show* in 1982 and served in that capacity until the show went off the air in 1989. (Courtesy of KPHO-TV.)

It was a Labor Day weekend tradition on Channel 5 to broadcast the nearly 24-hour *Jerry Lewis Telethon* to raise money for the Muscular Dystrophy Association. Many television stations throughout the country carried the program originating from Las Vegas, while others provided additional local studio segments showing local volunteers accepting call-in donations. (Courtesy of KPHO-TV.)

This three-camera studio production is a good example of the minimalist sets local TV stations used when producing a local program. Many shows were simply shot in front of a large curtain stretched in front of the studio walls. Called a cyclorama, this curtain could give the illusion of infinite space when lit properly. (Courtesy of KPHO-TV.)

KPHO-TV's news control room in the 1980s had separate video monitors for each camera, videotape recorder, film chain, or character generator. The separate control room allowed additional productions to take place while news was being originated from a different studio. (Courtesy of KPHO-TV.)

The movie *Star Wars* must have been fresh in the mind of the set designer when creating this KPHO-TV news set. In the 1980s, the anchor team included Stu Tracy, weather; Ken Coy, news; and Jon Brictson, sports. (Courtesy of KPHO-TV.)

The KPHO-TV news team in the late 1980s and early 1990s included Chris Cochran, Roger Downey, and Linda Turley. Channel 5 scheduled its evening news to air at 9:30 p.m. so it was not in direct competition with the 10:00 p.m. newscasts of the network-affiliated stations. (Courtesy of KPHO-TV.)

A news anchor team is the face of a local television station, which expends a great deal of time and money to find the right combination. Once those people are selected, they are heavily promoted to audiences, as exemplified by this ad for Diana Sullivan and Cary Pfeffer in the early 2000s. (Courtesy of KPHO-TV.)

Electronic newsgathering trucks allowed the news crew to send a signal directly back to the station, but there were limitations on the distance the signals could travel. However, after the launch of communications satellite Telstar in 1962, using satellites to transmit video became the norm. By the late 1980s, even local television news departments could afford to join the Space Age, and KPHO-TV, now a CBS affiliate, was justifiably proud of its new satellite uplink truck. (Courtesy of KPHO-TV.)

In 1971, KPHO-TV moved from its original location at 631 North First Avenue to new, ultra-modern studios at Indian School Road and Interstate 17. That was the station's home until 2013, when Channel 5 moved into the MAC America building, following the Meredith Corporation's purchase of Channel 3. KPHO-TV and KTVK-TV are now sister stations living in the same home. (Courtesy of KPHO-TV.)

Two

Wallace and Ladmo

Offering pictures, sound, movement, and music right in the family living room each day, it is no surprise that TV quickly captured the attention of children. Advertisers and television programmers also were quick to see the potential of providing programs just for kids. The programs were inexpensive to produce, relying on cartoon packages that had run in motion picture theaters a generation before. The cartoons were often introduced by a local announcer dressed as a clown or a superhero of the era. In the pre–*Sesame Street* days, the networks carried *Captain Kangaroo*, *The Howdy Doody Show*, and *Kukla, Fran and Ollie*. Local stations produced other kids' shows from nationally syndicated scripts, such as *Bozo the Clown* and *Romper Room*.

Phoenix had its own popular kids' show, which became the longest-running in the country with the same cast. *The Wallace and Ladmo Show* is remembered as a near-religious experience by nearly any kid who lived in Phoenix between 1954 and 1989. Indeed, several books have been written about the phenomenon.

Bill Thompson (Wallace) came to Phoenix in 1952, applied for a job at KPHO-TV, and was hired on the floor crew to operate the boom microphone for *The Goldust Charlie Show*. Host Ken Kennedy noticed Thompson's unique sense of humor and asked him to play the role of Wallace Snead in 1954. Within a year, Thompson was hosting his own kids' show—*It's Wallace*. After writing the comedy bits daily for the 60-minute show, Thompson soon decided he needed a foil for many of his jokes and recruited cameraman Ladimir Kwiatkowski to play the role of Ladmo.

By 1960, the year Pat McMahon joined the cast, the show had won its first Emmy award, Wallace and Ladmo were performing before stage audiences, and the program was reaching thousands of homes each day.

Thompson was always open to adding new talent to complement the show, such as Mike Condello, who made a major impact as music director, and in 1984, Cathy Dresbach and Ben Tyler from the Ajo Repertory Company.

Known throughout the years as *It's Wallace* (1954), *Wallace and Company* (1968) and, finally, *The Wallace and Ladmo Show* (1970), the show always received high ratings, and love and laughter from its viewers.

Built over three decades, *The Wallace and Ladmo Show* became an Arizona television institution. It began with Bill Thompson joining *The Goldust Charlie Show* on KPHO-TV in 1954 as Wallace Snead, Charlie's klutzy, unreliable nephew. From left to right are Bill Thompson as Wallace, guest actor/comedian Joe E. Brown, Ken Kennedy as Goldust Charlie, and unidentified "customers" of the fictitious general store, which catered mostly to children. (Courtesy of Robert T. Martin.)

The Wallace character gained enough popularity that Channel 5 chose Thompson to host his own cartoon show, *It's Wallace*, in January 1955. Thompson himself juggled all the writing and performing, including commercials. In January 1956, Thompson wanted to do two-man sketches, and asked one of his camera operators, Ladimir Kwiatowski, to join him. Viewers enjoyed Wallace playing "big brother" to Ladmo's child-like character. (Courtesy of the Wallace and Ladmo Foundation.)

By June 1956, *It's Wallace* had the highest ratings of any daytime show in Phoenix. Subsequently, celebrities began stopping by the show to promote their own careers. Shown here is Spike Jones, a popular musician, comedian, and bandleader, listening to Wallace play a tiny saxophone on *It's Wallace* in the late 1950s. Jones incorporated gunshots, whistles, and cowbells into satirical arrangements and lyrics of popular songs. (Courtesy of Robert T. Martin.)

Muhammad Ali appeared on *It's Wallace* four times over the years, first as Cassius Clay. Ali liked the show so much, he often had his driver drop him at the Channel 5 studio so he could watch the show as part of the audience, unannounced. The cast members were pleasantly surprised to see "the Greatest" sitting among the kids and parents. (Courtesy of the Wallace and Ladmo Foundation.)

Pat McMahon started his career in Phoenix as a KPHO-TV weather man / announcer. Daily, he left the newsroom to watch *It's Wallace* from the wings as a fan, impressed that the show never "talked down" to kids. One day in 1960, Thompson asked McMahon to join a skit; later, he joined the cast. Here, McMahon poses with "Wall and Lad" in an early Gerald costume. (Courtesy of the Wallace and Ladmo Foundation.)

The first two of many Emmy awards arrived in 1960 via presenters June Lockhart and Robert Stack, when *It's Wallace* won for "Best Local Entertainment Program" and "Best Children's Show." Lockhart had been known since 1958 as Timmy's mother, Ruth Martin, on *Lassie*, while Robert Stack was playing Eliot Ness on the *Untouchables*. From left to right are Thompson, Lockhart, Stack, and Kwiatkowski. (Courtesy of House of Broadcasting Inc.)

From left to right, Thompson, Edie Adams, Peter Falk, and Kwiatkowski pose with a 1962 Emmy Award. Adams was known for appearing on Ernie Kovac's television shows and pitching cigars in popular 1960s ads. Since 1957, Falk was known for roles in *The Untouchables* and *The Twilight Zone*. In 1964, Pat McMahon won his own Emmy for "Most Outstanding Local Television Personality." (Courtesy of House of Broadcasting Inc.)

Over the decades, Thompson, Kwiatkowski, and McMahon hosted the biggest celebrities of their time. Pictured with Wallace is Art Linkletter, the radio and television talk-show pioneer well known for *Kids Say the Darndest Things*, when he asked young children simple questions and received hilarious answers. Audiences could not get enough of little children describing their version of the world and their parents' actions and activities. (Courtesy of the Wallace and Ladmo Foundation.)

Pictured here in the 1960s, Wallace visits with Michael Landon of *Bonanza*, which aired on NBC from 1959 to 1973. Landon's television dad, Lorne Greene, also appeared on *It's Wallace* when McMahon's character Marshall Good presented Greene with his own law-enforcement badge. (Courtesy of the Wallace and Ladmo Foundation.)

Marshall Good was not the first person one would want to call for help in a threatening situation. The character himself explained why: he lived in the "Mild West," where instead of shoot-outs at the local saloon, there were panel discussions at the YMCA, and Good's personal code of conduct included, "Take cover, wait 'til it's over, beg for mercy if caught." (Courtesy of the Wallace and Ladmo Foundation.)

Wanting to reflect current trends, in 1964 Thompson and cast created Hub Kapp and the Wheels as a spoof of the British Invasion rock bands. Pat McMahon portrayed rock star Hub Kapp, wearing a pompadour wig accompanied by huge fake eyebrows and eyelashes. The comedy-sketch band was so popular with real teenagers, Capitol records recorded an album, and the group appeared on *The Steve Allen Show* six times. (Courtesy of the Wallace and Ladmo Foundation.)

At age 15, local musician Mike Condello (lower right) became musical director of *It's Wallace* and started appearing on the show wearing the outfit pictured here. A major player in the stellar trajectory of Hub Kapp, Condello wrote the music and played guitar as one of the Wheels. Of all the success associated with Wallace and Ladmo, Pat McMahon says the national fame of Hub Kapp and the Wheels amazes him most. (Courtesy of the Wallace and Ladmo Foundation.)

Liberace visits Ladmo, McMahon (as Aunt Maud), and Wallace. Besides telling the most morbid stories to children and misstating facts about the guests she was interviewing, Aunt Maud also flirted with most males. She mistakenly congratulated Liberace for the hit song *Autumn Leaves*, by Roger Williams. Liberace enjoyed his visit with the *It's Wallace* gang so much, he forgot his lines. (Courtesy of the Wallace and Ladmo Foundation.)

Pat McMahon's Aunt Maud was a deceptively sweet-looking elder who read from a book of morose tales that often left Ladmo in tears: mommy up and leaves the family, little children are never heard of again (for various reasons), and innocent people and animals are murdered (in various ways). After her 1961 debut, Aunt Maud was so popular, she was a regular character on *It's Wallace*. (Photograph by Sue Frazier.)

Gerald "the Brat" liked to blame others for his bad behavior. Just as Wallace had debuted as Goldust Charlie's nephew, Gerald was introduced as the snobby, spoiled, sneeringly superior nephew of Channel 5's general manager—and therefore, never punished. Kids loved to hate Gerald so much, he became one of McMahon's most well-known characters. This photograph shows Gerald in his later, most-recognized costume. (Courtesy of the Wallace and Ladmo Foundation.)

From left to right, Gerald, Ladmo, and Wallace visit Jack Williams, Arizona governor from 1967 to 1975. Williams established the executive budget, legislative budget committee, state personnel system, and the Department of Public Safety. He also served as Phoenix mayor for two terms in the late 1950s. Losing an eye to cancer as a teenager, after graduation in 1929, Williams began his career at KOY radio. (Courtesy of the Wallace and Ladmo Foundation.)

The *Wallace and Ladmo* cast created a long list of characters that became household names. Boffo (McMahon), a glum clown who disliked children, was "only in the biz to make a quick buck." When Boffo made an appearance at an amusement park, one of his activities planned for kids was to "put 'em in a shopping cart and shove 'em down a hill." (Courtesy of the Wallace and Ladmo Foundation.)

Captain Super represented a mutual love for comic books shared in real life by Lad, Thompson, and McMahon, who played the "superhero" who wore shoulder pads and charged for his services. In 1982, Wallace learned what Captain Super actually did: handled stalled cars and lost dogs for $5; television repair and light housekeeping for $10; and famine, plague, and locusts for $15; "no out of state checks accepted." (Photograph by Sue Frazier.)

Ladmo poses with producer/director Sharon Kelley in the KPHO-TV control room. Kelley directed about 1,700 episodes of *The Wallace and Ladmo Show* from 1982 until its last show. Arizonans said goodbye to Wallace, Ladmo, Pat McMahon's many characters, and others, on December 29, 1989—after a total of 9,000 episodes and 35 years on the air, five days a week. (Courtesy of Sharon L. Kelley.)

Steven Spielberg lived in Phoenix as a child. His 1997 letter to Bill Thompson, also known as Wallace, describes how Spielberg was transfixed by watching the antics of Wallace, Ladmo, Gerald, and other characters, and documents how the show inspired his career and depicts the love that parents and their kids had for the iconic children's show. (Courtesy of House of Broadcasting Inc.)

STEVEN SPIELBERG

December 17, 1997

Dear Wallace,

You don't mind if I call you "Wallace" do you? That's who you have been to me for most of my life.

I was actually somewhat starstruck reading your note on the inside cover of THE WALLACE AND LADMO SHOW book. Your show inspired me, made me laugh, made me think, and even raised my level of expectations whenever I looked around at things that could make me laugh.

Most of all, you guys made for a great childcare center. When my mom saw me and my three sisters parked in front of the TV set watching THE WALLACE AND LADMO SHOW, she knew, except for bathroom breaks, we wouldn't be anywhere else.

Hope life is treating you well.

All my best,

Steven

SS/km

Known throughout the years as *It's Wallace, Wallace and Company*, and finally, *The Wallace and Ladmo Show*, the format was simple and similar to many other kids' shows: cartoons, prize giveaways, and comedy sketches. It was the talent and creativity of the *Wallace and Ladmo* actors, writers, and musicians that made this particular show last 35 years, earning recognition as the longest-running show with the same cast and the longest-running daily show. In 2002, Ken and Laurie Easley and Randy and Theresa Murray purchased the building at 631 North First Avenue, where *Wallace and Ladmo* first aired. They restored the studio, and the remaining space houses several creative companies. Today, this mural on the First Studio building honors Bill Thompson, Pat McMahon (as Gerald), and Ladimir Kwiatkowski for entertaining countless children and adults with smart, wry, topical humor that captured the hearts of so many Arizonans for so many years. (Photograph by Ken Easley.)

Three
KTYL-TV, KVAR-TV, KTAR-TV, AND KPNX-TV CHANNEL 12

Arizona's second television station was licensed to broadcast in Mesa. In April 1953, Dwight "Red" Harkins, founder of the Harkins Theater Group, along with investors, put KTYL-TV on the air. The station operated on channel 12 and carried NBC programming. Like most stations of the time, KTYL-TV provided a significant amount of locally produced programs along with several hours of test pattern each day.

In 1955, Harkins sold the station to Pacific and Southern Broadcasting, owner of KTAR radio, and the KTYL call letters were changed to KVAR-TV. In 1959, the FCC allowed Channel 12's call letters to be changed to KTAR-TV and the studio to be moved to Central Avenue in Phoenix.

Karl Eller, owner of an outdoor billboard company, bought the radio and television stations in 1968 to form Combined Communications. Later, Eller went on to own Circle K convenience stores and eventually to have the business school at the University of Arizona named in his honor.

In 1979, Combined Communications was sold and broken up. With a final change of Channel 12's call letters to KPNX, the TV station and the *Arizona Republic* newspaper became part of media giant Gannett. What did not change was the NBC network affiliation. Even in 1994, when four of the five major commercial television stations in Phoenix swapped network affiliations, Channel 12 continued to carry NBC programming as it had from the station's inception.

NBC drew many viewers with programs such as *Bonanza*, *The Today Show*, and *The Huntley-Brinkley Report*. Channel 12 also had its own local lineup. The syndicated children's program *Romper Stomper* found a home in Phoenix, while KPNX had a news team headed by anchor Ray Thompson with reporters such as Gene McClain, Bill Stull, and Diane Kalas—one of the Valley's first female reporters and on-air hosts. By the 1980s, in keeping with the trends of the era, Channel 12's conventional newscast had morphed into *Action News*.

In 2011, KPNX-TV moved to its current location, *The Arizona Republic* building on East Van Buren Street, and later changed ownership to Tegna, a spin-off from the Gannett Corporation.

Channel 12 was the second television station in the Phoenix market. As it was licensed to Mesa, its first studio was located outside Phoenix city limits, on the Phoenix-Mesa Highway. The KTYL building no longer exists, as new ownership was able to convince the FCC to allow them to move the station to downtown Phoenix. (Courtesy of Ray Lindstrom.)

In 1953, Dwight "Red" Harkins, with other investors, put KTYL-TV on the air on channel 12. Harkins had become the youngest theater operator in the world, and was interested in bringing the best motion pictures to his audiences and the best talent to his television station. Here, he is shown interviewing Frank Sinatra on KTYL. Sinatra was one of the most popular musical artists of the 1940s and an Oscar-winning movie star by the 1950s. (Courtesy of Harkins Theatres.)

In the 1950s, KTAR radio and television were operating out of this modern building located at 1101 North Central Avenue in Phoenix. In the 1980s, the building was expanded and modernized to accommodate the Combined Communications offices, while KTAR radio, under new ownership, moved to a new location. (Courtesy of Ronald Conner.)

In the late 1950s, network television programming in color was brand new. KTAR-TV Channel 12 was an NBC affiliate and advertised RCA television sets as well as *The Jack Paar Show* on this billboard. RCA owned NBC at that time, and Jack Paar inherited *The Tonight Show* from Steve Allen. Allen's first broadcasting job was with KOY in Phoenix after he dropped out of Arizona Teacher's College, now known as Arizona State University. (Courtesy of Ray Lindstrom.)

In the very early days of television, most commercials were presented live in front of studio cameras each time the commercial was broadcast. Very soon, however, advertisers and stations began to record 30- and 60-second "spots" on location and on 16mm film, which could be replayed over and over. Here, Channel 12's cameraman Chuck Emmert, sales executive Ray Lindstrom, and an unidentified individual discuss a script for a Park Central shopping mall commercial. (Courtesy of Ray Lindstrom.)

Karl Eller grew up in Arizona and attended the University of Arizona. In the early 1960s, he purchased an outdoor advertising company. In 1968, he bought the KTAR stations and merged them into his outdoor advertising company to form Combined Communications Inc. In 1979, Eller sold Combined Communications to Gannett and moved on to head Columbia Pictures and later the Circle K convenience store organization. (Courtesy of House of Broadcasting Inc.)

Pictured here is the news team that KTAR-TV assembled to compete with news powerhouse Channel 10. Diane Kalas stands out as the only female reporter on the staff. (Courtesy of Ray Lindstrom.)

KTAR NEWS REPORTERS
(Clockwise from the top)

John Culea

Bill Redeler

Don Webster

Jack Frogrier

Diane Kalas

Gene McClain

Dick Gaither

Jack Clifford came to Phoenix after graduating from college, and soon found a sports anchor position at fledging KTVK-TV. He quickly moved into sales to bolster his meager income, which led him to Channel 12, and next to the position of general manager. Clifford is shown here on his first day in that new position at age 36. He quickly moved on to more high-profile positions, including CEO of the *Providence Journal*, and is probably best known as the founder of cable's Food Network. (Courtesy of Jack and Beverly Clifford.)

KTAR-TV had always been an NBC affiliate and benefitted from the popularity of *The Today Show*. Channel 12 developed its own version of a morning program, called *Today in Arizona*. Diane Kalas, a Channel 12 journalist and one of the first women on television in Arizona, hosted the program. (Courtesy of Ray Lindstrom.)

The Phoenix Suns premiered as an NBA expansion team in 1968 under general manager Jerry Colangelo. Karl Eller, who led the ownership group of the Suns, also owned Channel 12 through Combined Communications; therefore, the KTAR stations carried the Suns games. Here, sales executive Ray Lindstrom (second from left) and general manager Jack Clifford (far right) meet with executives from advertising sponsor Arizona Public Service. (Courtesy of Ray Lindstrom.)

Ray Thompson was Channel 12 news director and anchor in the 1960s and 1970s. Prior to moving to KTAR-TV, he worked as an anchor for Channel 3. Thompson was a stickler for integrity and accuracy during his long tenure in the Phoenix market and served to bring a high level of professionalism to the Channel 12 news department. (Courtesy of Phoebe Thompson.)

KTAR-TV's anchor team poses with Bill Denney's new Channel 12 Sports Trans Am. From left to right are Linda Alvarez and Kent Dana, news anchors; Bill Denney, sports anchor; and Dewey Hopper, weather reporter. Jerry Foster, a news helicopter pilot and reporter, stands on top of the car. Linda Alvarez anchored for Channel 12 from 1977 until 1985, when she moved to a station in Los Angeles. (Courtesy of Jerry Foster.)

As Phoenix was not so far by air from Hollywood, stars would often drop in the local television stations for publicity interviews. From left to right, Channel 12 anchor Vince Leonard is on the *Action News* set with comedian Red Skelton and Jerry Foster, the weekend weather guy. (Courtesy of Jerry Foster.)

Around the mid-1960s, KTAR-TV acquired this news van. Air conditioning was an add-on at that time, as were the blue emergency lights and the loudspeaker. (Courtesy of Ray Lindstrom.)

Channel 12's Jerry Foster directs videographer Chuck Emmert to get a shot of a Phoenix police officer at a crime scene. Foster was hired to be a helicopter pilot/reporter but soon was involved in other on-air jobs on the ground. (Courtesy of Jerry Foster.)

This Hughes "Jet" helicopter was considered the sports model of its generation. Channel 12's acquisition of this chopper instituted what became known as the "news helicopter wars" in the Phoenix television market, as each station promoted its helicopter as the first to get to the breaking news. (Courtesy of Jerry Foster.)

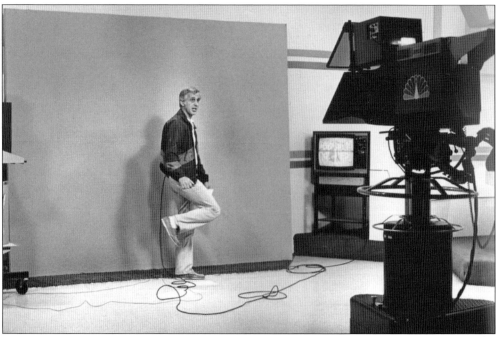

The weather is often reported on television in front of a "green screen." The flat green background allows the weather map and static graphics to be combined with an image of the weathercaster to create the illusion that the person is standing in front of a map. Here, weekend weather forecaster Jerry Foster delivers the forecast for the next day. (Courtesy of Jerry Foster.)

Dewey Hopper, KTAR-TV's umbrella-wielding weather man, had a more serious role hosting this Channel 12 telethon supporting Easter Seals in 1978. Channel 12, as well as most other local television stations, did much to support local and national charities during this time. (Courtesy of Ray Lindstrom.)

Channel 12's pilot Jerry Foster and videographer Chuck Emmert both took scuba diving lessons, which later came in handy when aiding the Maricopa Sheriff's Office in water rescues. Channel 12's news helicopter was used many times over the years as Foster stepped out of the journalist role to become a participant in the rescue and transportation of disaster victims. (Courtesy of Jerry Foster.)

While television helicopters were very useful as news-gathering tools, they often were also used for promotional purposes. In this picture, pilot Jerry Foster holds the door open for Santa and his helpers, who are delivering a chopper-load of toys for Christmas. (Courtesy of Jerry Foster.)

NBC televised Super Bowl XLIX from the University of Phoenix Stadium in 2015. Channel 12 anchors Lin Sue Cooney and Mark Curtis covered many of the surrounding events for local affiliate KPNX-TV. (Courtesy of Lin Sue Cooney.)

From left to right, KPNX-TV news anchors Caribe Devine, Mark Curtis, and Lin Sue Cooney appear on the Channel 12 news set for this promotional photograph. Most viewers select a local news channel because they feel an affinity for the anchor team. (Courtesy of Lin Sue Cooney.)

Four

KOOL-TV, KTSP-TV, AND KSAZ-TV CHANNEL 10

Young Tom Chauncey rolled into Arizona on a freight train. Desperate for cash, the 13-year-old found work as a bellhop at the Adams Hotel. He quickly ingratiated himself with the hotel's clients, the movers and shakers of Phoenix at the time. Chauncey next learned the jewelry business by working for Friedman's Jewelry down the street, later opening his own jewelry store in the Adams. The business thrived during the war years when many young airmen were stationed in Phoenix for training.

Gene Autry, a movie and radio star cowboy, came to Phoenix to learn to fly prior to entering the armed forces. He lived in the Adams and became friends with Chauncey. Autry wanted to own radio stations and asked Chauncey to be a local contact/manager so they could obtain an FCC license. Eventually, Autry's investment group sold the first station and bought radio station KOOL, which prospered under Chauncey's management.

In 1953, Autry and Chauncey applied for a permit for KOOL-TV to become the market's third television station. Unfortunately, KOY radio applied for the same channel, 10. Rather than undergo an expensive hearing before the FCC, the competing stations agreed to share the frequency, with each operating the station every other day. The agreement did not work, and KOY-TV was sold to KOOL-TV.

Chauncey had become friends with CBS executives, so KOOL-TV Channel 10 became the CBS affiliate. His network contacts brought to KOOL-TV its first studio cameras, donated DuMont hand-me-downs from WBBM-TV in Chicago. Television in the early days was not a profitable business and operation was expensive. However, under Chauncey's leadership and station manager Homer Lane's direction, KOOL-TV began to flourish.

Channel 10 was a leader with CBS programming, top-rated newscasts, and even a documentary unit that traveled worldwide to film. Then came change. Autry sold his part of the station, which forced Chauncey to do so, too. In 1982, the ownership went to Gulf United Broadcasting, and the station became KTSP-TV. It was sold again, and finally in 1994 became KSAZ-TV, owned by News Corp., making KSAZ a Fox network owned-and-operated station.

From its earliest days, Channel 10, along with KOOL-AM and FM, operated out of a large building at 511 West Adams Street. Originally the structure was an auto dealership thus having a large open space that was converted into a studio. A theater-type marque enabled KOOL to market station events and personalities to passersby on West Adams Street. News anchor Bill Close had the highest ratings during the 1960s and 1970s. (Courtesy of Neil A. Miller.)

KOOL-TV's co-owner Tom Chauncey (standing) meets with actor George Cleveland in the late 1950s. Note the coverage map in the background that shows KOOL-TV's signal reaching most of the populated areas of Arizona. (Courtesy of House of Broadcasting Inc.)

Gene Autry was a movie and radio star of the 1950s. The singing cowboy also was involved in ownership of radio and television stations. In addition to owning Golden West Broadcasting, with radio and television stations around the country, Autry partnered with Tom Chauncey to form KOOL-TV in Phoenix. KOOL-TV had a sister station in Tucson, KOLD-TV. (Courtesy of Ray Lindstrom.)

Election night was an important news event to be covered by television. In this photograph, newsman Jack Ware goes over some of the races while Bill Miller operates the DuMont studio camera. Miller went on to become news director at Channel 10, and later general manager of Channel 3. (Courtesy of Neil A. Miller.)

Like many early television stations, KOOL-TV sometimes covered live events outside of the studio. One wonders how the engineering staff got the very heavy DuMont camera to the top of this flimsy-looking scaffolding. While the cameraman on the left stands, the announcer on the right has made himself comfortable in a chair. (Courtesy of House of Broadcasting Inc.)

This photograph documents that KOOL-TV has covered local sports for more than a half-century. Here, the station's cameras cover the annual rivalry game between Arizona State University (ASU) and the University of Arizona in the 1950s at the former Goodwin Stadium. This stadium hosted its first football game in 1936, and the last in September 1958, after which ASU moved to Sun Devil Stadium. (Courtesy of House of Broadcasting Inc.)

Televising parades was popular in the 1950s and 1960s. KOOL radio and television used a vintage fire engine to promote the stations with the "KOOL is Hot" logo by entering it in those parades. (Courtesy of House of Broadcasting Inc.)

In the 1950s, a remote broadcast often meant that the engineering and production crews took the very large and heavy studio cameras out of the station and to the location to be televised. Here, KOOL-TV has its cameras out on the street to cover an event. (Courtesy of Neil A. Miller.)

This large bus was a mobile radio studio for KOOL as well as a great marketing device for the Channel 10 television station. In the 1950s and 1960s, going on location required transporting bulky, tube-driven electronic equipment. (Courtesy of Neil A. Miller.)

In the 1960s, actor Raymond Burr, the star of CBS's *Perry Mason*, visited with KOOL-TV staff. From left to right are Mary O'Hanlon, KOOL-TV traffic manager; Homer Lane, KOOL-TV general manager; Burr; and Ralph Painter, KOOL news. *Perry Mason* aired at 9:00 p.m. Thursday nights on Channel 10. (Courtesy of Ray Lindstrom.)

KOOL-TV program director Bob Martin, sales manager Les Lindvig, and station manager Homer Lane pay rapt attention to Hollywood actress Beverly Garland. In the late 1960s, Garland played the role of Barbara Harper on the CBS program My Three Sons. (Courtesy of Robert T. Martin.)

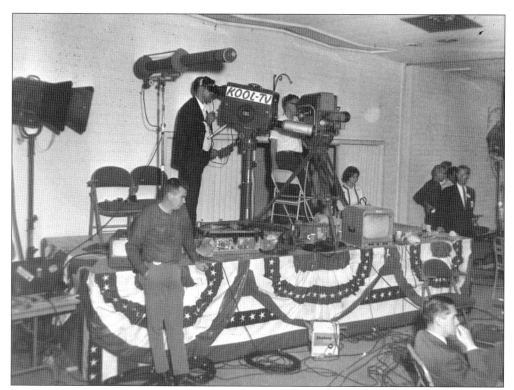

This KOOL-TV remote setup was needed to capture a speech or political rally in a downtown hotel ballroom. Like all remotes in the 1950s and 1960s, this one required much equipment and many people in order to broadcast the event. (Courtesy of Neil A. Miller.)

Later to become governor of California and president of the United States, Ronald Reagan was best known in the 1950s as the host of CBS's *General Electric Theatre*. In addition to hosting the show, Reagan traveled the country representing the General Electric company. Here, Reagan describes the advantages of the GE line of kitchen appliances in the KOOL-TV studios. (Courtesy of Robert T. Martin.)

Manny Garcia, the first Hispanic television news photographer in Phoenix, was hired away from Phoenix College in 1964 to establish their own news film unit at KOOL-TV. Nearly all remote news stories that required pictures were shot on black-and-white 16mm film, evidenced by the Arriflex camera that Garcia is lugging. (Courtesy of Manny Garcia.)

In the 1960s, news film ("Film at 10") was really film. News photographers covered the action with a 16mm silent film camera. Channel 10's chief photographer Manny Garcia is posing with his camera for a promotional newspaper photograph. (Courtesy of Neil A. Miller.)

Channel 10's early TV news van was often dubbed the "Rolling Stove" because of the in-cab engine, poor insulation, and ineffectual add-on air conditioning unit, which was not ideally suited to the 110-degree summer temperatures in Phoenix. (Courtesy of Manny Garcia.)

KOOL-TV is often given credit for having the first aerial newsgathering system in Phoenix. The McCulloch J-2 gyroplane shown here with pilot Jerry Foster was used for about a year before Channel 10 ordered a helicopter. A gyroplane does not have an engine connected to the rotor, and lift is only generated from the forward motion of the plane to turn the rotor. (Courtesy of Jerry Foster.)

KOOL's new McCulloch gyroplane is dwarfed by the commercial airliners of the day in this photograph at Phoenix Sky Harbor airport. The gyroplane did not prove to be reliable and was soon replaced by a more traditional helicopter. (Courtesy of Jerry Foster.)

America's most trusted man, CBS anchor Walter Cronkite, is shown here on a visit to Northern Arizona with KOOL-TV's pilot/reporter Jerry Foster in front of Channel 10's first helicopter. Cronkite was a frequent visitor to Phoenix. (Courtesy of Jerry Foster.)

Before television news choppers were customized with permanently mounted cameras, the pilot/reporter served as the videographer while flying the helicopter. Here, Jerry Foster operates an early Sony color camera for KOOL news with one hand while flying with the other. (Courtesy of Jerry Foster.)

67

Channel 10's documentary unit traveled the state and even the world, following stories that later appeared on the popular documentary series *Copper State Cavalcade*. Here, producer Bill Miller records an interview at the Lowell Observatory on Mars Hill in Flagstaff. The resulting program, *Long Eyes of Kitt Peak*, won a Peabody Award for the station. (Courtesy of Manny Garcia.)

After portable broadcast cameras and recorders became available in the early 1970s, Channel 10, as with most stations, quickly switched to the new, quicker, and less expensive format. Here, chief videographer Manny Garcia and producer Pam Stevenson record footage for a documentary. (Courtesy of Manny Garcia.)

In the early 1970s, Walter Cronkite provided a good lead-in to KOOL-TV's 6:00 p.m. news with Bill Close and Dave Nichols. Channel 10's evening newscast was the most watched in the Valley at that time. (Courtesy of Ray Lindstrom.)

The world of broadcast journalism seemed to shift on its axis in 1976 when Mary Jo West joined Bill Close on the KOOL news set. West was the first female news anchor in the market, and many of the old-line male anchors were not happy to share the news desk with a woman. However, Channel 10's ratings continued as the best in the market. (Courtesy of Neil A. Miller.)

In 1979, Jerry Foster (helicopter pilot/reporter), Bill Denney (sports anchor), and Kent Dana (weekend anchor) all left KOOL-TV for similar positions at Channel 12. In an interview years later, Dana revealed that he doubled his salary when he moved to Channel 12. Dana served as a news anchor for 25 years before he moved to a similar position with Channel 5. (Courtesy of Jerry Foster.)

Channel 10's newsroom was where reporters developed contacts and wrote stories, and producers worked out the lineup of the day's stories. It had become common in the 1980s for local television stations to use their newsrooms as auxiliary sets during the evening newscasts. From left to right are Diane Ryan, Ron Hoon, and Fred Kalil. (Photograph by Neil A. Miller.)

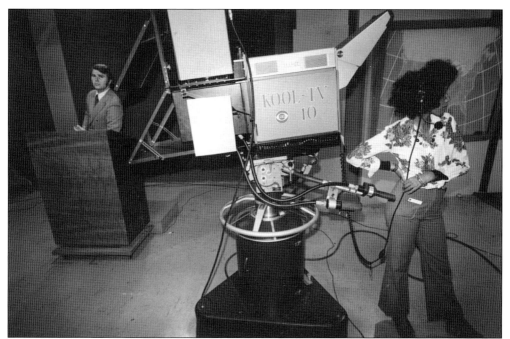

RCA had developed its third generation of color studio cameras by the 1970s, when KOOL-TV purchased the TK-44A version. The box-like structure on top of the lens is a video monitor that displays a script that is reflected by a mirror into a two-way glass positioned in front of the lens. This teleprompter allows the talent to read from a script while looking directly into the camera lens. (Photograph by Neil A. Miller.)

Mary Jo West, Phoenix's first female prime-time news anchor, went on to CBS in New York, then back to Phoenix to anchor on Channel 3. She later served as station manager for the City of Phoenix cable channel. In this photograph, she is seen talking with Walter Cronkite at Arizona State University's annual Cronkite Luncheon in the early 1990s. (Photograph by Bob Rink.)

On May 28, 1982, Joe Billie Gwin (left) walked into KOOL-TV's Washington Street news studio, pulled out a gun, and fired a shot into the ceiling. He grabbed production assistant Luis Villa, put the gun to his head, and demanded that his manifesto be read over the air. (Courtesy of KOOL-TV.)

After several hours of negotiations, KOOL interrupted its regular evening programming to broadcast anchor Bill Close (left) reading the rambling statement Gwin had prepared—all with a gun pointed at him while he sat next to Gwin. After the broadcast, Gwin surrendered to police and was incarcerated for several years. As a result, security was tightened at television stations across the country. (Courtesy of KOOL-TV.)

Bill Leverton has just filmed an aerial view for a story. With Jerry Foster at the controls of the helicopter, Leverton feels that he has sure footing to step onto the top of the news van. (Courtesy of Jerry Foster.)

In an unannounced move, co-owner Gene Autry sold his 50 percent of KOOL-TV to Gulf United Broadcasting. Tom Chauncey was then forced to sell his portion of the radio and television stations. Chauncey's last day of owning and managing the stations came in 1982. His loyal employees are seen here bidding him farewell. Channel 10 was to become KTSP-TV and later—under Fox ownership—KSAZ-TV. (Photograph by Neil A. Miller.)

Much of what the news anchor says on air is actually composed behind the camera. News producers are responsible for selecting, organizing, and often writing the news content. From left to right, reporter Norma Cancio works with Channel 10 producers Doug Drew and Phil Alvidrez. (Courtesy of Douglas Drew.)

News producers Dennis O'Neil (left) and Doug Drew are hard at work polishing the scripts that anchors Mary Jo West and Bill Close will read on the 6:00 p.m. newscast. Computers of the 1980s looked a bit different than they do today, but were capable of sending the script directly to the teleprompter. (Courtesy of Douglas Drew.)

Channel 10's *Celebrate Arizona* promotion took the evening newscast to multiple locations around Arizona. In this photograph, Dave Patterson and Deborah Pyburn prepare to broadcast from an observatory. (Photograph by Neil A. Miller.)

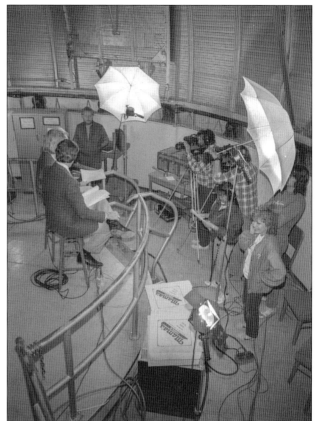

It takes a lot of equipment to originate the evening news from a remote location. Here, anchors Dave Patterson and Deborah Pyburn prepare to broadcast the evening news from the Painted Desert. (Courtesy of Bill and Bonnie Leverton.)

In addition to all of the technology required for a remote broadcast, *Celebrate Arizona* required a crew of thousands—well, maybe only 13, but still a significant number. Here, the journalists and crew pose for a group photograph in Monument Valley, Arizona. (Courtesy of Bill and Bonnie Leverton.)

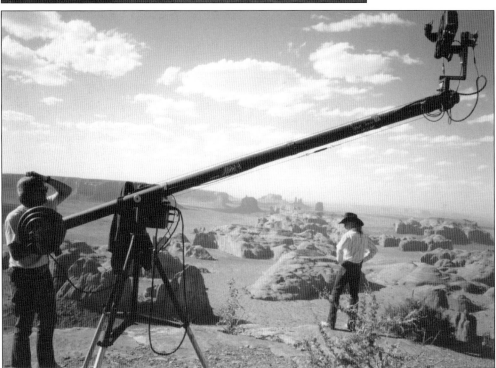

A camera mounted on a boom allowed KOOL-TV to get very dramatic camera movements for a shoot in the spectacular scenery of Monument Valley. Prior to self-contained satellite communication trucks, live broadcasts from such remote locations would have been impossible. (Courtesy of Bill and Bonnie Leverton.)

Before *Celebrate Arizona* took viewers around the state for remote newscasts, Bill and Bonnie Leverton were uncovering "feel-good" stories in distant locations with their *On the Arizona Road* series. Here, Leverton reports from the edge of the Grand Canyon. (Courtesy of Bill and Bonnie Leverton.)

The Channel 10 crew prepares to shoot a segment on the beach at Rocky Point, Sonora, Mexico. Reporter Bill Leverton reads over the script, while the local kids find the production lights more interesting. (Courtesy of Bill and Bonnie Leverton.)

Sports anchor Jude LaCava and news anchor John Hook are longtime Channel 10 news personalities. They are shown here in 2001 broadcasting live from a remote location. (Courtesy of House of Broadcasting Inc.)

All television stations must be ready to cover special events as they occur. Pope John Paul II visited Phoenix in 1987. In this photograph, he stands on the balcony of St. Mary's Basilica in downtown Phoenix. Local television stations pooled equipment and personnel to provide coverage of this historic event. Maurie Helle, production manager of KOOL-TV, served as television pool producer/director for this major undertaking. (Courtesy of Ray Lindstrom.)

Five

KTVK-TV Channel 3

Earnest McFarland has been the only Arizonan to serve significant positions in all three branches of government: as a US senator, Arizona governor, and the chief justice of the Arizona Supreme Court. He also founded a television station—KTVK Channel 3—the fourth television station serving Phoenix when it signed on air in 1955. When KTVK signed an ABC network–affiliation contract, it stripped KPHO-TV of its last network programming source.

McFarland became interested in the new medium of television when serving as a member of the US Senate's Communications Subcommittee. At first, his station operated out of a specially designed studio building on North Sixteenth Street. Initially, most of its news and local programs did not attract large audiences, nor did the ABC network programming at that time.

After McFarland died in 1985, his daughter Jewell and her husband, Delbert Lewis, took over station ownership. In 1986, Bill Miller left Channel 10 as news director to become general manager of KTVK and brought with him several Channel 10 executives to change the fortunes of Channel 3. News director Phil Alvidrez quickly expanded news programming under the brand NewsChannel 3. Marketing efforts under the brands of Arizona's Family and The Place with More Stuff allowed the station to garner some of the highest ratings in the market.

In 1994, network affiliations were switched between four of the five major television stations in Phoenix. After more than 40 years, KTVK-TV lost its ABC affiliation and became an independent station. Against all odds, the station survived and even thrived on syndicated programming and local news.

MAC America (the Lewis family media outlets) sold Channel 3 to the A.H. Belo Corporation in 1999, thus ending local ownership of the major television outlets in Phoenix. All were now under corporate ownership and managed from afar as part of larger media groups.

In 2013, Belo sold the station to the Meredith Corporation, which also owned KPHO-TV Channel 5. KPHO moved into the larger Channel 3 building and began operating independent KTVK-TV as well, according to the FCC rule that limits ownership to only two television stations in the same market.

Ernest McFarland was a US senator, governor of Arizona, and chief justice of the Arizona Supreme Court. In addition, he established KTVK-TV in Phoenix on channel 3 in 1955 under the entity of Arizona Television Company. (Courtesy of Ray Lindstrom.)

Delbert and Jewell Lewis assumed ownership of KTVK-TV in 1985 after the death of Senator McFarland. Jewell was McFarland's daughter, thus keeping the ownership in "Arizona's Family," which became the station's brand. Arizona Television Company, which had been renamed MAC in honor of McFarland, had become the largest family-owned media organization in the country when it was sold to Belo in 1999. (Courtesy of House of Broadcasting Inc.)

News videographers were on the job day and night. Here, KTVK-TV's Jamie Ontiveros shoots 16mm film of a fire scene in Phoenix for later use on the 6:00 p.m. news. The big disadvantage of film was that it had to be taken back to the station and processed before being edited and loaded onto a projector for broadcast—a very time-consuming process. (Courtesy of Carol Lynde.)

Channel 3 photographer Sean Nicolai captures the action on 16mm film in this photograph. By the 1960s, local television stations were shooting film with a magnetic stripe on the side. This allowed sound to be recorded along with the picture. (Courtesy of Carol Lynde.)

In the early days of television, local stations often wanted to originate programs from outside the studio. Many times, old bread or dairy delivery vans were repurposed to serve as television remote trucks. Homemade camera platforms were often attached to the top of the vehicles to obtain pictures from above the action. This Channel 3 truck has not been used in a very long time. (Courtesy of KTVK-TV.)

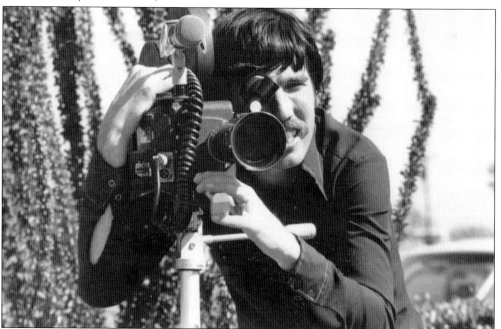

Photographer Sean Nicolai captures the action on 16mm film. In the 1970s, news departments transitioned from 16mm film to 3/4" videotape, shot with a video minicam. The advantage was increased speed from shooting to broadcast, as hours of film processing time were eliminated. (Courtesy of KTVK-TV.)

In the 1970s, before news stations used microwave-delivered reports from remote locations, the lowly family station wagon served as a news car to transport the reporter, videographer, and equipment to the site. A flashy paint job helped promote the station. (Courtesy of Carol Lynde.)

Channel 3's news team in the late 1970s included, from left to right, Ray Thompson, Dave Nichols, Jim Howl, Evelyn Thompson, and Tom Schoendist. Prior to this, Thompson was the first news anchor on Channel 3, then was a longtime news anchor and news director for KTAR-TV. Dave Nichols previously anchored for KOOL-TV news. (Courtesy of Carol Lynde.)

Carol Lynde went to work for KTVK-TV in 1976 as the first female videographer in the Phoenix market. The job was not an easy one in those days as the videographer had to carry the camera on one shoulder and the video cassette recorder on the other. Generally, videographers also wore a very heavy battery belt to power the equipment. (Courtesy of Carol Lynde.)

"It's breaking news," "We are going live," and "Quick, get that shot"—videographers Jim Pulliam and Carol Lynde are on the job for KTVK-TV news during a breaking news event. (Courtesy of Carol Lynde.)

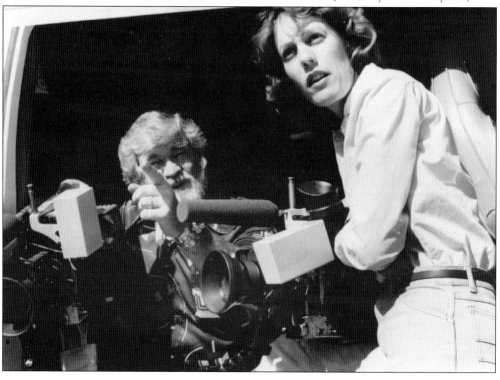

"I'm cold and wet, and my boots are full of water, but the camera and recorder are safe," perhaps Carol Lynde laments. "Maybe the life of a television videographer isn't going to be as glamorous as I thought." (Courtesy of Carol Lynde.)

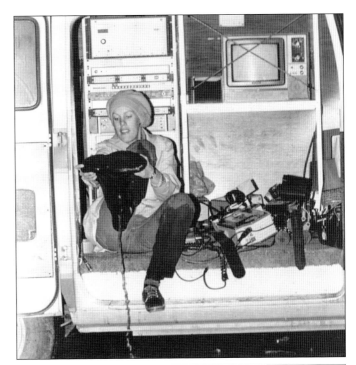

In the 1970s and 1980s, many stations customized tradesman's vans in order to provide less expensive remote units for news departments. The addition of a microwave dish and generator allowed the unit to transmit a signal back to the station within a limited geographic area. (Courtesy of Carol Lynde.)

By the 1980s, the helicopter had become an important news-gathering tool. In the early 2000s, there were five television news helicopters flying in Phoenix. Channel 3's news helicopter circles over University Avenue and the Arizona State University campus in Tempe. Often, Bruce Haffner could be found at the controls while reporting from the air. (Courtesy of Carol Lynde.)

KTVK-TV, billing itself as Arizona's Family, has long been involved in civic events and fundraising for charity. It also has taken pains to let audiences know that they should remember the channel number. (Courtesy of KTVK-TV.)

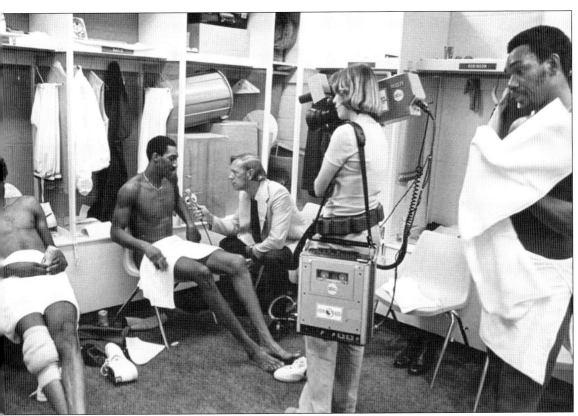

In the mid-1970s, there was a great amount of controversy surrounding the right of a female journalist to enter a male dressing room to cover a sports team or player. In 1978, a US district court ruled that, under the 14th Amendment, a woman had an equal right to pursue her career. That same year, Channel 3 videographer Carol Lynde broke the gender barrier by becoming the first female journalist in Arizona to take her camera inside the Phoenix Suns' locker room. Shown here, Lynde records an interview conducted by Channel 3 sports reporter Tom Schoendist. (Courtesy of Carol Lynde.)

After losing its ABC affiliation in 1994, KTVK-TV was able to contract with CNN for national and international news coverage. To fill other former network hours, general manager Bill Miller and program director Sue Schwartz turned to syndicated shows such as *Jeopardy* and *The Oprah Winfrey Show*. A large amount of time was given to news director Phil Alvidrez for news programming. Here, anchor Jodi Applegate prepares for a morning news broadcast. (Courtesy of KTVK-TV.)

Longtime anchor Patti Kirkpatrick (formerly of KPNX-TV) interviews a history re-enactor at a local school visit. Kirkpatrick was paired with Heidi Fogelsong on Channel 3's evening news for several years. It was very unusual at this time for a television station to have two female news persons anchoring on the same newscast. (Courtesy of KTVK-TV.)

Anchor and reporter Patti Kirkpatrick (center) visited many local schools throughout Arizona in order to interview teachers for an evening news segment. The public relations value of the school visits was important in generating ratings for the local news programs. (Courtesy of KTVK-TV.)

Channel 3 lost its ABC affiliation in 1994 and had to scramble to find new programming. When a morning news program named *Good Morning Arizona* featuring Jodi Applegate was developed, it was decided to put the show on the air quickly in order to beat Channel 10's planned morning show. A hastily made set of plywood and sawhorses emphasized the "under construction" nature of the project. *Good Morning Arizona* was the top-rated morning show for years afterward. (Courtesy of Phil Alvidrez.)

Good Morning Arizona was unique as a morning news program for its time. It included a great deal of entertainment and public service information as well as hard news. On-air personalities were an important factor in the success of the program. Here, Dan Davis (back row, center) and Brad Perry (back row, far right) pose with guests from the Susan G. Komen breast cancer awareness program. (Courtesy of KTVK-TV.)

Satellite communication made it possible for news crews to originate live stories from almost anywhere in the state—or the world, for that matter. Here, Channel 3 KTVK-TV is on-site in Prescott, Arizona. (Courtesy of KTVK-TV.)

Once Channel 3 obtained a satellite uplink truck, the station had the ability to broadcast live from almost any location. The decision was made to originate live news programs from around the state in a promotional campaign named Arizona, U.S.A. Here, KTVK-TV's Frank Camacho and Heidi Foglesong prepare to broadcast the evening news live from the rim of the Grand Canyon. (Courtesy of KTVK-TV.)

It is hard to miss which channel owns this satellite truck. The large dish on the back of the truck directs the signal to a geostationary communication satellite some 22,300 miles above the equator. That signal is then retransmitted back to the television station for broadcast to local audiences. (Courtesy of KTVK-TV.)

Local television news had become very important in the late 1980s and early 1990s. New news-gathering tools such as the minicam and satellite communication had made it possible to originate live programs from almost anywhere in the world. During this time, KTVK-TV devoted significant resources to filling many hours a day with news. The station even sent reporters and equipment to cover stories in different parts of the world. (Courtesy of Lisa Schnebly Heidinger.)

A Channel 3 videographer is ready to get that shot. When on a remote location, the best angle may require the videographer to crawl to the top of the truck. A two-way radio allows the director to give instructions regarding the type of shot that will be needed for the live program. (Courtesy of KTVK-TV.)

It takes more personnel than just those seen on camera to make a location shot possible. In this photograph, the technical crew operates camera controls, video switchers, and audio consoles in the KTVK-TV remote truck. (Courtesy of KTVK-TV.)

Live television remotes require a great amount of preparation on the part of the engineering department as well as the news department. Lights, cameras, headsets, and microphones all require cables to be connected and tested before the broadcast, and the cameras must be color-balanced (using the chart seen here at center) to produce a realistic picture. (Courtesy of Lisa Schnebly Heidinger.)

"The scenery in Northern Arizona looks beautiful on TV, but do we have four-wheel-drive on this truck? I don't think that AAA will come to rescue us out here," might have been a comment from the news crew. From left to right are Tom Heidinger and Brian Nelles. (Courtesy of Lisa Schnebly Heidinger.)

"When I was studying journalism, no one said that my job would include driving a huge motorhome across the Arizona desert," Jodi Applegate may have thought. KTVK's 1994 Arizona, U.S.A. campaign, which originated news from the far corners of the state, required complex logistics. Applegate is driving the vehicle to reach the remote site where she will appear on-air. (Courtesy of Sharon Harnden.)

While technology certainly makes it easier to tell the news story in a timely manner, it still takes a reporter to track down the details and verify their accuracy. Here, KTVK-TV reporter Jana Wallis condenses the facts into a coherent description of the event. Very soon, she will use her handwritten script to aid her in delivering her report on camera. (Courtesy of Sharon Harnden.)

Longtime reporter Steve Bodinet provides a standup report from a remote location. Such distant coverage was made possible by the development of news-gathering technology. However, it takes the hard work of the journalist to ferret out facts and organize them in a coherent manner. Only then can the public understand the world as reported on the nightly news. (Courtesy of Sharon Harnden.)

With the adoption of any new technology comes a learning curve. While the helicopter had been around for decades and used as a news-gathering platform by a few stations for a number of years, it only became pervasive in the 1990s. Soon, all five Phoenix news copters would be seen hovering in close proximity over the same news event. At 12:46 p.m. on July 27, 2007, a tragedy occurred over Steele Indian School Park in downtown Phoenix. All five television helicopters, as well as a Phoenix police helicopter, were following a Phoenix police car chase. Several of the pilots were broadcasting live as they described the events, communicated on the radio, maneuvered to get good camera shots, and attempted to stay clear of each other. Aircraft operated by KTVK-TV and KNXV-TV collided, and journalists Scott Bowerbank and James Cox of Channel 3 and Richard Krolak and Craig Smith of Channel 15 lost their lives. One consequence of this tragedy was that the stations now share helicopter footage rather than operate their own aircraft. Another change was that a pilot must now concentrate only on the complex task of flying while a second person serves as the reporter. (Courtesy of KTVK-TV.)

Six

KAET-TV Channel 8

Some thought that television could open a window to the art world and bring great painting, sculpture, music, and drama to the masses. Educators envisioned television as the great master teacher that would close the gap between the underfunded school districts of the inner cities and the posh suburban campuses of the more affluent. This was especially a concern during the space race, when the need for better instruction in science and math became apparent.

Frieda Hennock, an FCC commissioner, lobbied for and obtained a reservation of 25 percent of all television channels for education when the FCC released its allocation of channels in 1952. Many school districts, state education departments, and universities quickly obtained licenses to operate educational television stations, soon discovering that constructing and operating television stations is very expensive.

Arizona State University applied for a license through the Board of Regents, and KAET-TV Channel 8 signed on the air in 1961. Like many educational stations, KAET limped along with only subsistence funding. First, KAET offices were located in double-wide house trailers parked next to ASU's engineering building, which housed space for a studio. Channel 12, then KTAR-TV, provided castoff camera equipment to allow Channel 8 to begin operations. KAET even used a retuned KTAR-TV auxiliary transmitter, antenna, and transmitter building in its early days.

By the mid-1960s, most realized that educational television was not fulfilling its potential due to lack of funding needed to produce programming and provide an interconnected network. Congress passed the Public Broadcasting Act of 1967, which provided the mechanism to launch the Public Broadcast Service (PBS) network, and by 1969, public stations across America were teaching five-year-old kids the alphabet and how to count under the tutelage of Big Bird and the Cookie Monster on *Sesame Street*.

Today, a better funded Channel 8, located in downtown Phoenix, is operated by ASU's Walter Cronkite School of Journalism and Mass Communication. It provides five channels of quality programs to most Arizona homes. Popular programs include *Downton Abbey*, *PBS NewsHour*, *This Old House*, *Antiques Roadshow*, and *Frontline*, to name a few.

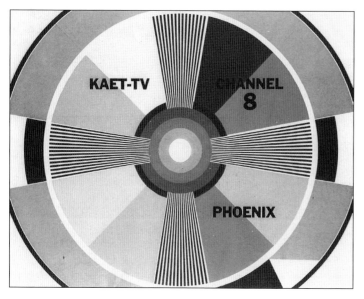

In the very early days of television, most stations broadcast test patterns when programs were not scheduled, because they did not have enough programs to air round-the-clock. Originally designed to allow engineers to tweak the output of cameras and transmitters, the patterns soon became familiar to all early viewers. KAET-TV used a version of the test pattern as its station identification. (Courtesy of KAET-TV.)

Pledge drives became a main source of funding for most public television stations. After Richard Bell, director of ASU's radio-television bureau, applied for the license in 1958, Bob Ellis took over management of the station in 1961. Ellis served in that position for nearly 30 years and is seen here hosting a pledge drive. Audiences watching KAET-TV volunteers take call-in pledges would have a hard time missing the telephone number. (Courtesy of KAET-TV.)

Public stations often were at the forefront of hiring women and minorities for jobs behind the camera as well as in front of it. Marla Coleman is seen here operating an audio console at KAET-TV. (Courtesy of KAET-TV.)

General manager Bob Ellis is shown with some of the pledge-drive volunteers who answered the telephones. While commercial stations could sell time to advertisers, public broadcasting had to find other ways to raise money to buy programming rights and operate the stations. Some early programs could be obtained at low cost from National Educational Television, which mailed filmed programs from station to station. (Courtesy of KAET-TV.)

A young Michael Grant (far right) was the first host of *Horizon*. Here, he confers with the science editor and the executive producer of the program. *Horizon* first went on-air in 1981. (Courtesy of KAET-TV.)

Local public affairs program *Horizon*, hosted first by Michael Grant and later by Ted Simonds in 2008, became a signature production for KAET-TV. *Horizon* airs five days a week. (Courtesy of KAET-TV.)

Phyllis Palacio, an early co-anchor on *Horizon*, tracks down a news source. Palacio was a broadcasting student at the Walter Cronkite School of Journalism and Mass Communication while working at Channel 8. (Courtesy of KAET-TV.)

The KAET logo in the 1970s cleverly combined the call letters with the channel 8 number. The station ID became an important promotional device to solidify the station channel number with the call letters in the mind of the viewer. (Courtesy of KAET-TV.)

Director Tony Schmitz sets up a shot for KAET-TV's first nationally distributed series on American Indian artists. The programs were shot on 16mm film by Don Cirillo in the early 1970s. (Courtesy of KAET-TV.)

Poet and songwriter Rod McKuhen records narration for the *American Indian Artists* series in the KAET-TV audio facility. (Courtesy of KAET-TV.)

KAET-TV was proud of its Rocky Mountain Southwest Emmy award issued by the Phoenix chapter of the National Academy of Television Arts and Sciences. Over the years, the station has been very competitive in winning awards for locally produced programs. Channel 8 won its first Emmy in 1964. (Courtesy of KAET-TV.)

Smaller battery-powered cameras allowed KAET-TV to shoot footage on location. Here, Rick Balch composes a shot for the *Exploring Lake Powell* program. (Courtesy of KAET-TV.)

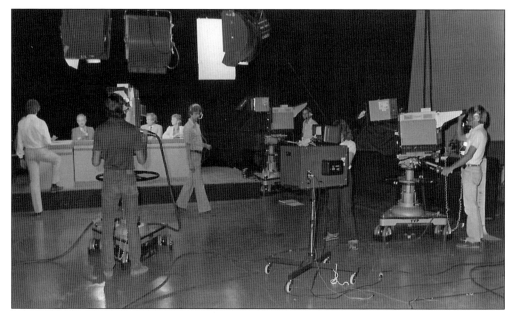

A larger studio, equipped with three RCA TK-44 color cameras, allowed KAET-TV to originate public affairs discussion programs during the mid-1970s. Multiple cameras were necessary for live programs. As the signal of one camera was being broadcast, other cameras could prepare for the next shots. (Courtesy of KAET-TV.)

Public television stations have long relied on locally produced interview programs to inform the public about important issues that are often overlooked by commercial television news outlets. Channel 8 has provided information on many important topics over the decades. (Courtesy of KAET-TV.)

Arizona resident and popular author, columnist, and humorist Erma Bombeck appeared not only on ABC's *Good Morning America*, but also on KAET-TV. Bombeck was a longtime resident of Paradise Valley, Arizona, and was beloved locally and nationally for her take on suburban home life, from the mid-1960s until her death in 1996. (Courtesy of KAET-TV.)

Michael Grant and former Arizona governor Jane D. Hull pose on the *Horizon* set after a broadcast. Elected officials used *Horizon* often over the decades to explain public policy to citizens of the state. (Courtesy of KAET-TV.)

PBS was among the first to use satellite communications to distribute a regular program schedule to affiliated stations in 1976. The technology of the time, labeled C-band, required very large receiver dishes. These dishes were located adjacent to Stauffer Hall, KAET-TV's former home on the ASU Tempe campus. (Courtesy of KAET-TV.)

In the 1980s and 1990s, KAET-TV originated instructional programs for college credit from university classrooms. The programs were distributed to subscribers through private channels. ASU produced its first telecourse for credit in 1959. The instructional program in beginning Spanish was recorded in makeshift studios in the engineering building and broadcast by KPHO-TV Channel 5 as a public service. (Courtesy of Elizabeth A. Craft.)

KAET-TV partnered with Seattle public television station KCTS-TV to produce *Over Arizona* in the mid-1990s. The program coupled panoramic helicopter footage with symphonic music to bring majestic Arizona scenery, such as this found in Monument Valley, to the television screen. (Courtesy of KAET-TV.)

In 1983, program director Chuck Allen (later general manager) produced *The Operation*, which was seen worldwide. *The Operation* was an open-heart procedure performed on live television by Dr. Ted Detrick at St. Joseph's Hospital and Medical Center. (Courtesy of KAET-TV.)

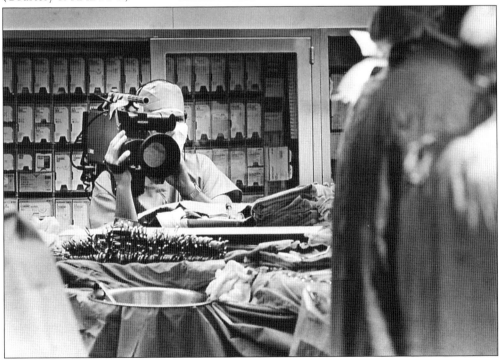

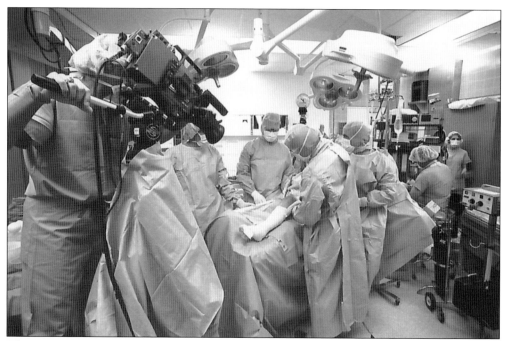

The Operation was so successful that KAET-TV decided to produce two more live medical programs. In this photograph, doctors prepare to replace a patient's knee joint. In 1989, KAET-TV produced a third operation program, in which hip replacement surgery is performed. (Courtesy of KAET-TV.)

In 2001, the Arizona Diamondbacks won the World Series. Carol Lynde and David Majure covered the action at what was then Bank One Ballpark for KAET-TV. Even public broadcasting was excited when the World Series came to town. (Courtesy of Carol Lynde.)

Seven

KNXV-TV Channel 15

After a long television career in the Northeast, Edwin Cooperstein moved to Arizona to establish KNXV-TV Channel 15. Although the FCC issued the construction permit in 1975, it took nearly five years for the station to sign on the air. As an independent station on a UHF channel, before cable television in the city—and in a highly competitive market—KNXV experienced an economic slough. Subscription television offering first-run movies at night was thought to be the answer, but the viewers never materialized and the provider, ON-TV, was out of business in Phoenix by 1983. KNXV-TV then was required to operate as a truly independent station.

E.W. Scripps, a large TV station group owner, bought KNXV-TV in 1984, obtained better-syndicated programs, and won the rights to local sports teams such as the Phoenix Suns. In 1986, the station became an affiliate of the new Fox network to provide a Phoenix outlet for the late-night *Joan Rivers Show*. As Fox grew to produce other programs, the lineup improved.

A network affiliation shakeup in 1994 sent Fox to channel 10 and CBS to channel 5. KNXV might have once again become an independent station, but Scripps negotiated a contract with ABC, and KTVK-TV was left without a network. Scripps had the negotiating power, owning the top-rated ABC affiliated stations in Detroit and Cleveland.

KNXV-TV built a modern facility near Forty-Fourth and Van Buren Streets, with much of the ground floor devoted to a newsroom/studio combination. With an emphasis on news, the station quickly become a ratings force in the Phoenix market and won many Emmy awards for its investigative reports. KNXV added live news trucks, staff, and a news helicopter. Later, however, in order to cut costs, the station added multimedia journalists to the mix: reporters were required to report, shoot, and edit their own news stories. The station eventually joined a local helicopter news service in order to share news video with other stations in the market as an additional cost-saving measure.

By the 2000s, the station was looking very much like the typical ABC affiliate in any large city.

KNXV-TV signed on the air as an independent in the Phoenix market on September 9, 1979. Its first programming schedules were based on early off-network syndicated situation comedies and old movies. Originally, the station was in the area of Fortieth Street and Roeser Road in a former warehouse. The station had moved to a group of three buildings in South Phoenix by the time it received the ABC affiliation contract. (Courtesy of KNXV-TV.)

While Channel 15 ran sitcoms and movies during the daytime, it was the Phoenix home of ON-TV in prime time. The subscription service required a set-top receiver to decode the signal and a monthly payment from each home using the programming service. (Courtesy of KNXV-TV.)

KNXV-TV has been housed in three different facilities since its inception in 1979. Pictured here is the master control room in the station's second location, on South Thirty-Third Place. A technician selects audio and video from several programming sources prior to sending the signal via microwave to the transmitter on South Mountain. (Courtesy of KNXV-TV.)

In an early KNXV-TV studio, the anchors and crew prepare for an evening broadcast. This photograph captures the typical bustle of activity on the news set before going on-air, including interaction with producers and the floor crew. Scripps purchased KNXV in 1985 and signed on with the new Fox network in 1986. In 1994, the station moved to a new location and received its ABC affiliation. (Courtesy of KNXV-TV.)

KNXV had developed its first local news to debut in 1994; it was a fast-paced newscast branded "No Chit Chat, More News," to match the style of the new Fox network. But it never broadcast news as a Fox affiliate, as it changed affiliation to ABC. The "Al's World" convertible seen here once carried KNXV reporter Al Feinberg around the state to cover feature stories. (Courtesy of KNXV-TV.)

When KNXV introduced news as part of the station's programming, Marc Bailey and Darya Folsom were among its first news anchors. They are shown here preparing for a newscast. (Courtesy of KNXV-TV.)

On location, a KNXV news crew is working to send a remote broadcast back to the station via microwave signal. The dish must be high in order to send the signal above trees and buildings that might otherwise block its path. A class of grade-school students watches the action. (Courtesy of KNXV-TV.)

This picture documents television technology of the past. Reel-to-reel video tape recorders and 3/4" videotape machines dominate the equipment racks. Today, this would be a wall of computer servers. (Courtesy of KNXV-TV.)

Grand-opening day at Channel 15's new home near Forty-Fourth and Van Buren Streets was January 1, 1999. Members of the general public are touring the new building. (Courtesy of KNXV-TV.)

Having a Humvee as a television news van not only stood out, it meant that ABC 15 could go just about anywhere with this four-wheel-drive, all-terrain vehicle. (Courtesy of KNXV-TV.)

Channel 15's new, larger studios on Forty-Fourth Street allowed the inclusion of audiences for town hall public affairs programs. This program was hosted by former ABC News anchor Peter Jennings and featured a distinguished panel including Hugh Downs and Gov. Janet Napolitano. (Courtesy of KNXV-TV.)

This is a tighter shot of the same town hall that the home audience saw, which eliminates all the backstage activity that the in-studio audience could see. (Courtesy of KNXV-TV.)

Occasionally, a story requires taking the satellite truck out of the Valley of the Sun to cover it. Here, reporter Mark Lodato and his photographer are in Northern Arizona experiencing something that they rarely have to contend with in Phoenix: snow. (Courtesy of Mark Lodato.)

KNXV-TV's two-story newsroom is located adjacent to the news studio. The "Investigators" sign indicates Channel 15's emphasis on in-depth, investigative news coverage. The Investigators have won local and national awards, including a 2008 Peabody Award for investigating breaches in Phoenix Sky Harbor's security system. Today, the station promotes its ABC 15 Investigators as the largest investigative team in the state. (Courtesy of KNXV-TV.)

Eight
OTHER PHOENIX TELEVISION STATIONS

While the focus of this book is primarily the traditional, full-power television stations that have operated in Phoenix from the very early days of the medium, many other TV stations have come and gone over the past 70 years.

Several of note include those that targeted specific audiences. Some were programmed to reach the large Spanish-speaking population in the Phoenix metropolitan area. Other stations were programmed for home shopping, one has had a religious format, and one station is an independent licensed to Prescott, Arizona. Three stations were carrying programming from the short-lived networks—the WB, UPN, and Pax. These stations were on the less-desirable UHF frequencies, and some were low power, or even translators extending the signals of other stations, and thus had very small audiences, with a few exceptions.

The oldest of these stations is KPAZ-TV, which signed on the air in 1967 as a Spanish language station, but is now owned by the Trinity Broadcasting Network. Trinity provides religious programming to the Valley on channel 21. The second Spanish language television station to go on air was KTVW-TV Channel 33 in 1979. This Univision station operated out of a mobile home in South Phoenix in its early days as a Spanish International Network affiliate. Due to its success, in later years it moved to a modern glass and rammed-earth structure in a beautiful desert landscape at Southern Avenue and Thirty-Second Street. It often earns the highest ratings for its programs and has won numerous Emmy Awards for its local programming.

Perhaps the most visible of the English-language independent stations is KAZT-TV, which is marketed as AZ-TV Channel 7. The FCC first assigned channel 7 to Prescott and a license to KUSK-TV in 1982. KUSK operated as an independent station serving Northern Arizona until it filed for bankruptcy. Jack Londen purchased the station in 2002 and began serving the Phoenix market through a translator that local TV sets could pick up as channel 7. By programming off-network syndicated series, local sports programs, and a daily live talk show called *Arizona Daily Mix*, AZ-TV has built a respectable audience.

KPAZ-TV signed on the air in 1967 as the first full-power, independent, UHF (channel 21) station in Phoenix, operating in a facility at Tower Plaza Mall. It could not compete economically with KPHO-TV and filed for bankruptcy in 1969. The station was sold to the Glad Tidings Church, which changed the format to include religious programming. In 1975, KPAZ-TV moved to a new facility on East McDowell Road. (Photograph by John E. Craft.)

Channel 21's early programming was bilingual—a mixture of English and Spanish—including bullfights from Mexico and jai alai games. Currently, KPAZ-TV is owned by Trinity Broadcasting Network and carries religious programming, as well as children's programs, on digital channels. (Courtesy of Ray Lindstrom.)

When Univision Arizona KTVW 33 went on the air in 1979, a double-wide mobile home served as its studio and offices. By 1982, the station had moved into an actual structure on Southern Avenue and Thirtieth Street. That facility was replaced by an ultra-modern rammed earth and glass building across the street. (Courtesy of Univision Arizona.)

The Spanish International Network (SIN) was the forerunner to Univision and can trace its beginnings to 1962, when a Spanish language station in San Antonio provided programming to two additional stations. In 1987, SIN was sold to a partnership of Hallmark Cards and Televisa to form Univision. This 1982 photograph shows the Univision Arizona KTVW-TV satellite dish in front of the Channel 33 studio on Southern Avenue and Thirtieth Street in South Phoenix. (Courtesy of Univision Arizona.)

In 1992, Channel 33 began broadcasting *Teleradio 33*, a live weekday morning show. Laura Wright, the host of *Phoenix de Noche*, is seen here interviewing a guest with Jose Ronstadt, host of *Teleradio 33* and general manager of Univision Arizona. (Courtesy of Univision Arizona.)

Channel 33 has a full news department and sends reporters out on location to interview newsmakers who would be of interest to Spanish-speaking audiences. In this photograph, videographer and producer Brian Acosta is recording an interview with Laura Wright, host of *Phoenix de Noche*. The subjects are members of the international music group Pandora, who were in town in 1994 to promote their tour "Ilegal." (Courtesy of Univision Arizona.)

This 1990 photograph was taken in the studio during a broadcast of *Teleradio 33*. From left to right are Jose Ronstadt, host of *Teleradio 33* and general manager of KTVW-33; the guest, wrestler Super Pepe; cohost ? Isabel; and floor manager Gina Santiago. Masked and costumed wrestlers were quite popular with Hispanic audiences. (Courtesy of Univision Arizona.)

KTVW-TV, with a sister station in Tucson and several translators, covers the state with news broadcasts delivered in Spanish. Channel 33's news often gains higher ratings than the local English language newscasts. While this might be due to the English audiences being split between several stations competing for the same viewers, credit also must be given to KTVW's professional news team. (Photograph by John E. Craft.)

Visitors approaching the studios of Univision Arizona in South Phoenix are now greeted with a vast, carefully landscaped desert garden, which forms a perfect setting for its award-winning home. Channel 33 has come a long way since its days working out of a mobile home across the street. (Photograph by John E. Craft.)

This is the front entrance of the Univision Arizona KTVW 33 office and studio on Southern Avenue and Thirtieth Street in South Phoenix. The 35,000-square-foot facility was built in 2001 by Chanen Construction Company and designed by Swaback Partners, the architectural heir to famed designer Frank Lloyd Wright. In 2001, this project was honored as one of the 18 greatest architectural achievements in Arizona by the American Institute of Architects. (Photograph by John E. Craft.)

Phoenix-based Londen Companies bought KUSK-TV, renovated the Prescott facility, renamed the station KAZ-TV, and opened a main office and studio at Forty-Fourth Street and Camelback Road in Phoenix. Depending on where viewers lived, they could find KAZ-TV on channels 23, 27, or 57, and Cox cable channel 13, and the station could be seen across Central and Northern Arizona. Recently, KAZ-TV was able to negotiate carriage on cable channel 7 from Cox. (Courtesy of KAZ-TV.)

Dodie Londen poses with Ron Bergamo. Londen was vice chairwoman of Londen Companies and chairwoman of its Londen Media Group division. She chose Bergamo to lead KAZ-TV as the station's first vice president and general manager based on Bergamo's success when leading then KSAZ-TV Channel 10. Bergamo had also managed stations in Tucson as well as Texas and Kansas. (Courtesy of KAZ-TV.)

KAZ-TV engineers needed to import the broadcast signal from Prescott to the Phoenix market in high-quality digital, as well as deliver live content from a new production facility in Phoenix for rebroadcast in Prescott (the city of the station's license) and then back to the Phoenix market for transmission on the Cox cable system. This is the production control room in Phoenix. (Courtesy of KAZ-TV.)

Pat McMahon's talk show was among the first local programs aired on KAZ-TV. From left to right are Pat McMahon, Dodie Londen, and Jack Londen. McMahon was well-known in Arizona for playing many characters over the years on *The Wallace and Ladmo Show* and being a highly-rated radio talk show host on KTAR radio. (Courtesy of KAZ-TV.)

Former *Arizona Highways* publisher Win Holden is interviewed by host Pat McMahon in the early days of his talk show on KAZ-TV. (Courtesy of KAZ-TV.)

Jack Londen, Sam Steiger, and Dodie Londen chat at KAZ-TV's grand opening. One of the station's first talk show hosts, Steiger was known for his role in Arizona politics and for serving five terms in the US House of Representatives. He also gained publicity during his career for his strong, independent, Western attitude. As his wife described him, "You either liked him or you didn't—there wasn't much middle ground." (Courtesy of KAZ-TV.)

Local civic events are important, and in order to be an accepted part of a small community, a local television station should cover those events. In this photograph, KUSK-TV on-air hosts describe a Prescott parade for the television audience. The bed of a pickup truck makes a good platform to raise the anchor desk above the action. (Courtesy of KAZ-TV.)

When KAZ-TV debuted in Phoenix, in addition to local talk shows, other local programs included college sports on weekends. The station had agreements with ASU and the University of Arizona to broadcast sports events; this billboard promotes University of Arizona basketball and football under the previous call letters KUSK-TV. (Courtesy of KAZ-TV.)

The House of Broadcasting Inc.

The House of Broadcasting Inc. is a 501c3 nonprofit organization dedicated to preserving the history of radio and television in Phoenix. It was founded in 1997 by Mary Morrison, a media buyer who worked with most of the Phoenix broadcasting stations during the course of her career. She noticed that, as new ownership and employees came to the stations, much of the history was in danger of being lost. With the help of Jack Clifford, founder of the Food Network; Tom Chauncey II, media attorney and son of the founder of Channel 10; Art Brooks, president of the Arizona Broadcasters Association; and the Phoenix chapter of American Women in Radio and Television, Morrison established a small museum first located in a display window at Christown Mall in West Phoenix. Later, the museum moved to Fifth Avenue in Scottsdale. Currently, the House of Broadcasting has exhibits in the Cronkite Gallery at the Walter Cronkite School of Journalism and Mass Communication in downtown Phoenix and at the Arizona Historical Society museum in Tempe.

In 2012, the Rocky Mountain Southwest chapter of the National Academy of Television Arts and Sciences recognized the House of Broadcasting with a Governor's Award for outstanding service to the profession and the community. In addition to operating a broadcasting museum, the organization raises funds needed to award scholarships to students in the Walter Cronkite School of Journalism and Mass Communication and the Hugh Downs School of Human Communication—both at Arizona State University—and the Eller College of Management at the University of Arizona. In order to fund its programs, the organization has sponsored an annual Christmas concert, awarded recognition to outstanding media leaders at banquets, sponsored celebrity golf tournaments, held book signings, sold cookbooks with the recipes of local television personalities, and has accepted contributions from private donors.

As the technology and the content of electronic mass communications move forward at an ever-increasing pace, the House of Broadcasting Inc. looks forward to preserving the past.

Discover Thousands of Local History Books
Featuring Millions of Vintage Images

Arcadia Publishing, the leading local history publisher in the United States, is committed to making history accessible and meaningful through publishing books that celebrate and preserve the heritage of America's people and places.

Find more books like this at
www.arcadiapublishing.com

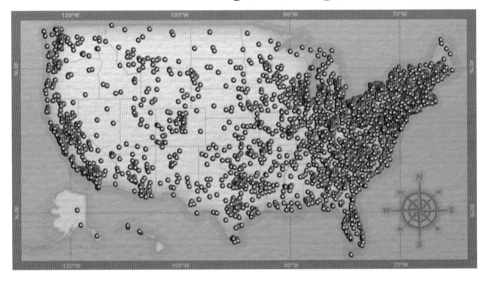

Search for your hometown history, your old stomping grounds, and even your favorite sports team.

Consistent with our mission to preserve history on a local level, this book was printed in South Carolina on American-made paper and manufactured entirely in the United States. Products carrying the accredited Forest Stewardship Council (FSC) label are printed on 100 percent FSC-certified paper.